Derived Functors and
Sheaf Cohomology

Contemporary Mathematics and Its Applications: Monographs, Expositions and Lecture Notes

Print ISSN: 2591-7668
Online ISSN: 2591-7676

This series aims to inspire new curriculum and integrate current research into texts. Its aims and main scope are to publish:

- Cutting-edge Research Monographs
- Mathematical Plums
- Innovative Textbooks for capstone (special topics) undergraduate and graduate level courses
- Surveys on recent emergence of new topics in pure and applied mathematics
- Advanced undergraduate and graduate level textbooks that may initiate new directions and new courses within mathematics and applied mathematics curriculum
- Books emerging from important conferences and special occasions
- Lecture Notes on advanced topics

Monographs and textbooks on topics of interdisciplinary or cross-disciplinary interest are particularly suitable for the series.

Published

Contemporary Mathematics and Its Applications
Monographs, Expositions and Lecture Notes

Vol. **2**

Derived Functors and Sheaf Cohomology

Ugo Bruzzo
International School for Advanced Studies, Trieste, Italy
Universidade Federal da Paraíba , Brazil

Beatriz Graña Otero
Universidad de Salamanca, Spain

World Scientific

NEW JERSEY · LONDON · SINGAPORE · BEIJING · SHANGHAI · HONG KONG · TAIPEI · CHENNAI · TOKYO

Published by

World Scientific Publishing Co. Pte. Ltd.

5 Toh Tuck Link, Singapore 596224

USA office: 27 Warren Street, Suite 401-402, Hackensack, NJ 07601

UK office: 57 Shelton Street, Covent Garden, London WC2H 9HE

Library of Congress Cataloging-in-Publication Data

Names: Bruzzo, U. (Ugo), author. | Graña Otero, Beatriz, author.

Title: Derived functors and sheaf cohomology / Ugo Bruzzo (International School for Advanced
Studies, Trieste, Italy) and Beatriz Graña Otero (Universidad de Salamanca, Spain).

Description: Hackensack, NJ : World Scientific, [2020] | Series: Contemporary mathematics and
its applications : monographs, expositions and lecture notes, 2591-7668 ; vol. 2 |
Includes bibliographical references and index.

Identifiers: LCCN 2019035012 | ISBN 9789811207280 (hardcover) |
ISBN 9789811207297 (ebook)

Subjects: LCSH: Functor theory. | Sheaf theory. | Spectral sequences (Mathematics)

Classification: LCC QA169 .B78 2020 | DDC 514/.23--dc23

LC record available at https://lccn.loc.gov/2019035012

British Library Cataloguing-in-Publication Data

A catalogue record for this book is available from the British Library.

For any available supplementary material, please visit
https://www.worldscientific.com/worldscibooks/10.1142/11473#t=suppl

Desk Editors: V. Vishnu Mohan/Kwong Lai Fun

Typeset by Stallion Press
Email: enquiries@stallionpress.com

Sine doctina vita est quasi mortis imago
Dionysius Cato, 4th. c. AD

Preface

These notes are meant to be an introduction to the machinery that in homological algebra is called "derived functors", with a particular emphasis on sheaf cohomology; also, the examples in Chapter 5 are mainly from sheaf theory. Basically, all material presented here has been taught in MSc or PhD courses over the years. As a whole, it may be the basis of a one-semester course in sheaf theory, very much taken from a homological algebraic perspective.

A very distant precursor of Chapter 3 and the section on Čech cohomology (Section 4.1) are the notes of a course given in Florence in 1984 by our friend and collaborator Daniel Hernández Ruipérez. We thank him for in some sense introducing us to sheaves. Part of this material was already included in [3]. We also thank Daniel for the discussions.

We thank Claudio L.S. Rava for providing Example 3.25 and for the useful discussions, and Vitantonio Peragine for his very careful and critical reading of the manuscript and for giving us precious feedback. Chapter 5 includes parts of the notes taken by Alessandro Giacchetto during a course given at Scuola Internazionale di Studi Superiori Avanzati (SISSA), Trieste. We thank him for passing us his notes. More generally, we thank all the students at SISSA and elsewhere (in particular, Florianópolis and São Paulo), who were taught parts of this material by one of us, for their patience and their feedback.

This book was mainly written during the visits that we made, in different times and combinations, to SISSA, Universidad de Salamanca, University of Pennsylvania, Universidade Federal de Santa Catarina (Florianópolis), Universidade de São Paulo, Universidade Federal da Paraíba (João Pessoa) and Pontificia Universidad Javeriana at Bogotá. We thank all those institutions for their support and hospitality. Additional support was provided by Istituto Nazionale di Fisica Nucleare, Programma di Ricerca Scientifica di Rilevante Interesse Nazionale "Geometria delle varietà algebriche", Gruppo Nazionale per le Strutture Algebriche, Geometriche e loro Applicazioni dell'Istituto Nazionale di Alta Matematica (INdAM-GNSAGA), Pontificia Universidad Javeriana at Bogotá, Instituto de Física Fundamental y Matemáticas in Salamanca, the Spanish MEC through the research project MTM2013-45935-P, Universidad de Salamanca through Programa XIII, Conselho Nacional de Pesquisa (CNPq), Fundação de Amparo à Pesquisa do Estado de São Paulo (FAPESP grant 2017/22091-9).

About the Authors

Ugo Bruzzo is Professor of Geometry at SISSA, International School for Advanced Studies, Trieste, Italy, where he has been working since 1994. He has held long-term visiting positions at the University of Pennsylvania in Philadelphia, USA, Universidade Federal de Santa Catarina, Florianópolis, Brazil and Universidade Federal da Paraíba, João Pessoa, Brazil, and short-term visiting positions at Universidad de Salamanca, Spain, Université de Lille, France, Université Paris VI, France and Tata Institute for Fundamental Research in Mumbai, India. He specializes in algebraic and differential geometry and mathematical physics, in particular, moduli spaces of sheaves, moduli of quiver representations, and applications to topological quantum field theory. He already published two monographs, *The Geometry of Supermanifolds* (Reidel, 1991) and *Fourier–Mukai Transforms in Geometry and Mathematical Physics* (Birkhäuser, 2009). He has been the Editor-in-Chief of the *Journal of Geometry and Physics* from 2003 to 2017.

Beatriz Graña Otero received her PhD from Universidad Complutense of Madrid in 2003 and has been working as Assistant Professor and then Associate Professor of Algebra at Universidad de Salamanca, Spain since 1999. She has held visiting positions at SISSA in Trieste, Italy, University of Pennsylvania in Philadelphia, USA and Tata Institute for Fundamental Research in Mumbai, India and was Associate Professor at Pontificia Universidad Javeriana, Bogotá from 2012 to 2017. She wrote several papers in algebraic geometry and enjoyed a long and extended collaboration with Ugo Bruzzo.

Contents

Introduction

The word "homology" comes from the classical Greek words ὁμός (meaning "same") and λόγος (meaning "discourse" among other things), so it might be rendered as "discourse about sameness", or "study of the similarities". This term is widely used in science (biology, geology, etc.) and also in mathematics. True to their name, for instance, homological theories associated with topological spaces may be regarded as a way to find similarity patterns among spaces, or, in more precise terms, to associate *global invariants* to them. In some loose sense, the homology of a topological space detects the "holes" in the space; however, one should think of "holes" in all dimensions. So, a circle has a one-dimensional hole, but in the same way, a 2-sphere, which cannot be shrunk to a point, has a "two-dimensional hole". The way homology takes account of that is by associating a "chain" to every subspace and considering chains that have no boundary — called cycles — and chains that are boundaries; then, cycles modulo boundaries represent the "holes". Considering duals of chains in a suitable sense, one gets a cohomology theory, which somehow describes more refined invariants. Cohomology was first precisely distinguished from homology in Whitney's work [66].

The foundations of this theory were laid by Henri Poincaré in his *Analysis situs* [51], although some traces of similar ideas were already contained in the work of previous mathematicians, e.g., Euler's polyhedron formula $v - e + f = 2$, where v, e and f are the number of vertices, edges and faces of a closed polyhedron, respectively. Poincaré's *Analysis situs* developed into modern algebraic topology.

1

Another cohomology theory which associates global invariants with geometric spaces is *de Rham cohomology;* to be more precise, de Rham cohomology is defined for differentiable manifolds, although it turns out to be a very deep invariant, which actually only depends on the homotopy type of the differentiable manifold.[a]

A powerful generalization is provided by the cohomology theories that one can associate with *sheaves.* The main ones are *sheaf cohomology* and *Čech cohomology,* which we shall introduce and study in some detail in Chapter 4. The plus of the sheaf-theoretic approach is that one can describe not only properties of a topological space but also of the "functions" that one defines on it; for instance, two schemes X and Y may have the same underlying topological space, but may be equipped with different "sheaves of functions" (structure sheaves), and the resulting cohomologies may be different. A rather detailed and intriguing story of the first introduction of the notion of sheaf by Leray, very much from the viewpoint of algebraic topology, is given by Houzel's introductory chapter to Kashiwara–Schapira's book [36] (in French).

Homology and cohomology also appear in other contexts, even purely algebraic. A typical instance is provided by the *extension problems.* Suppose that we are working with a category where it makes sense to consider structures like

$$0 \to a \xrightarrow{i} b \xrightarrow{p} c \to 0, \tag{0.1}$$

where i is an injective morphism, p is surjective, and the image of i coincides with the kernel of p (this is an *exact sequence*). Then the following problem arises: if a and c are given, can we find a third object b with morphisms i and p that make up an exact sequence like (0.1)? If such triples exist, how can we classify them up to a suitable notion of equivalence? It turns out that these questions are answered by suitable cohomology groups, which contain the obstruction to find a triple, and classify the possible triples when the obstruction vanishes. This is the case with the extension problem in many categories, such as groups, modules, Lie algebras, associative algebras, Lie algebroids, and more.

Homology theories are ubiquitous in mathematics, and each of them can be defined in an *ad hoc* way. It is therefore quite important to be able to define and study them in a unified way. Such an approach is often provided by the theory of *derived functors.* Let us hint on the meaning

---

[a]Historically, de Rham cohomology was first introduced in de Rham's thesis [14].

of this terminology. First, a category is a set, or, more generally, a class, whose elements are called *objects*, and a set or class whose elements are called *arrows* and may be thought of as maps between objects. A *functor* between two categories $\mathfrak{A}$ and $\mathfrak{B}$ is a correspondence which maps objects of $\mathfrak{A}$ to objects of $\mathfrak{B}$ and arrows between objects of $\mathfrak{A}$ to arrows between the corresponding objects of $\mathfrak{B}$, compatibly with the composition of arrows. If it makes sense to talk about exact sequences in the categories $\mathfrak{A}$ and $\mathfrak{B}$, we may ask if a functor $F : \mathfrak{A} \to \mathfrak{B}$ maps exact sequences to exact sequences. If this is the case, we say that F is *exact*. However, it may happen that by applying F to an exact sequence like (0.1), one gets

$$0 \to F(a) \xrightarrow{F(i)} F(b) \xrightarrow{F(p)} F(c),$$

i.e., the morphism $F(p)$ may fail to be surjective; when this happens, we say that F is *left exact*. The (*right*) *derived functors* of F measure the failure of F to be exact; instead of an exact sequence

$$0 \to F(a) \to F(b) \to F(c) \to 0,$$

we rather get

$$0 \to F(a) \to F(b) \to F(c)$$
$$\to R^1 F(a) \to R^1 F(b) \to R^1 F(c) \to R^2 F(a) \to \cdots .$$

These objects $R^i F$, the right derived functors of F, are a kind of generalized cohomology groups, and indeed, cohomology theories can often be cast in this framework.

Another central argument in this book is the theory of *spectral sequences*. A spectral sequence arises whenever a complex has a filtration which is compatible with the differential of the complex. A (cohomology) complex is an object K in an abelian category that can be written as

$$K = \bigoplus_{n \in \mathbb{Z}} K^n$$

and is equipped with a morphism $d \colon K \to K$ (the differential of the complex) such that $d(K^n) \subset K^{n+1}$ and $d^2 = 0$. A filtration of K is a collection of subobjects

$$\cdots \subset F_{p+1} \subset F_p \subset \cdots \subset F_1 \subset F_0 = K,$$

such that

$$F_p = \bigoplus_{n \in \mathbb{Z}} F_p^n \quad \text{with } F_p^n \subset K^n \quad \text{for all } p$$

and $d(F_p) \subset F_p$. This structure generates a sequence of cohomology complexes such that the cohomology of a complex is isomorphic to the next complex. Under some assumptions, this sequence "converges" in a suitable sense, and the complexes obtained at the various steps, called the *pages* of the spectral sequence, yield a kind of approximation to the limit complex. So, this procedure can provide a technique for computing cohomology groups by successive approximations (there is dual theory for homology). Also spectral sequences are due to the creativeness of Jean Leray.

A few words about the structure of this book. Chapter 1 introduces the basic notions in category theory and homological algebra that will be needed in the sequel. In the part on homological algebra, emphasis is given to cohomology as opposed to homology, however the difference is purely formal, as at this level, one can swap between the two theories just reversing the directions of the differentials.

Chapter 2 introduces the main characters of this book, namely, derived functors. Our purpose in writing these notes is not that of being encyclopedic; rather, we hope to transmit to the reader the gist of what the theory of derived functors is. So, we mostly treat *left exact functors* and their *right derived functors* (another reason for this choice is that the theory of right derived functors is the one which readily applies to sheaves). We only give a cursory introduction to the "dual" theory of right exact functors and their left derived functors in Section 2.7 (however, some of the notions we introduce in this section will pop up in the exercises and will make a final appearance in Section 5.6.8 on the Künneth spectral sequences).

As our main example of a derived functor will be sheaf cohomology, in Chapter 3, we give a rather self-contained introduction to presheaves and sheaves from a purely topological perspective. Chapter 4 introduces first Čech and then sheaf cohomology, offering a careful comparison between the two cohomologies, culminating in the proofs of their equivalence for paracompact spaces or for quasi-coherent sheaves on noetherian separated schemes. In particular, in view of the last mentioned result, Section 4.5 provides a basic introduction to the theory of schemes.

In Chapter 5, we develop the theory of the spectral sequences associated with a filtration of a differential complex. We describe the main results (to simplify the treatment, we mostly consider "first quadrant spectral sequences") and discuss several specific spectral sequences, often regarding them as special cases of the spectral sequence associated with a hyper-derived functor.

Chapter 6 is a kind of epilogue: the natural outcome of the theory of derived functors is the notion of derived category, and indeed, the same mathematician who created the modern approach to sheaf cohomology, i.e., Grothendieck, introduced, with his student J. L. Verdier, the notion of derived category. So, this chapter offers a brief but hopefully clear introduction to this beautiful notion, referring the reader to the literature for further developments. The hope here is to entice the readers who will be so patient to follow us till this point to continue their study by approaching this elegant and interesting theory.

While we claim no originality in any sense, we would like to stress that the discussion of some issues is more detailed than what is usually found in the literature. For instance, this is the case for the relation between Čech and sheaf cohomology for paracompact spaces, which is discussed by showing that in this case, Čech cohomology is a universal δ-functor (Theorem 4.31). Whenever possible, we have tried to develop the theory in a unified manner; for instance, most examples of spectral sequences are obtained from the spectral sequences associated with a hyperderived functor.

All examples are mostly worked out in detail, while exercises are left to the reader (up to some hints). Both the examples and the exercises scattered through the chapters make essential part of the text. Each chapter ends with some additional exercises.

Chapter 1

Basic Notions

In this chapter, we establish the main terminology and language that we shall use throughout. The main ingredients will be *category theory* and *homological algebra*. Our introduction will be of course quite cursory; appropriate bibliography will be given in the text.

1.1. Category Theory

While we shall not be using that much of category theory, that will still be part of our basic language. Therefore, in this section, we offer a quick introduction to this theory. Not all concepts will be defined, as this section aims more at developing some basic intuition rather than providing a full, consistent introduction. For more comprehensive treatments, the reader may refer, e.g., to [17, 42, 43].

1.1.1. Categories and morphisms

Definition 1.1. A category $\mathfrak{C}$ is

- a class $\mathrm{Ob}\,\mathfrak{C}$ whose elements are called *objects*;
- for every $a, a' \in \mathrm{Ob}\,\mathfrak{C}$ a class $\mathrm{Hom}_{\mathfrak{C}}(a, a')$, possibly empty, whose elements are called *morphisms* (intuitively, we may think of morphisms as "arrows" between objects);

- for every $a, a', a'' \in \text{Ob}\,\mathfrak{C}$ an associative binary operation

$$\text{Hom}_{\mathfrak{C}}(a, a') \times \text{Hom}_{\mathfrak{C}}(a', a'') \to \text{Hom}_{\mathfrak{C}}(a, a'')$$

$$(f, g) \mapsto g \circ f$$

with the property that for every object a in $\mathfrak{C}$ there exists a morphism $\text{id}_a : a \to a$ such that for all $f \in \text{Hom}_{\mathfrak{C}}(a, a')$ one has $\text{id}_{a'} \circ f = f \circ \text{id}_a = f$.

The last property implies that every object a in $\mathfrak{C}$ has exactly one identity morphism id_a.

A category $\mathfrak{C}$ is said to be *small* if $\text{Ob}\,\mathfrak{C}$ is a set, and for every $a, a' \in \text{Ob}\,\mathfrak{C}$, the class $\text{Hom}_{\mathfrak{C}}(a, a')$ is a set as well; *locally small* when only the second condition holds.

Example 1.2. The class $\mathfrak{Set}$ of all sets, with morphisms given by the maps between sets, forms a *large* (i.e., not small) category. A given set S is a small (discrete) category whose objects are the elements of S, and all morphisms are the identity morphism.[a] If S has a preorder, we may form a small category whose objects are the elements of S and the morphisms are the order relations between the objects.

Given a category $\mathfrak{C}$, a *subcategory* $\mathfrak{S}$ of $\mathfrak{C}$ is a category whose class of objects is a subclass of $\text{Ob}\,\mathfrak{C}$, and for all objects a, a' of $\mathfrak{S}$, $\text{Hom}_{\mathfrak{S}}(a, a')$ is a subclass of $\text{Hom}_{\mathfrak{C}}(a, a')$. Moreover, $\mathfrak{S}$ is said to be a *full subcategory* of $\mathfrak{C}$ if $\text{Hom}_{\mathfrak{S}}(a, a') = \text{Hom}_{\mathfrak{C}}(a, a')$.

The opposite category $\mathfrak{C}^{\text{op}}$ of a category $\mathfrak{C}$ is the category with the same objects as $\mathfrak{C}$, while for every $a, a' \in \text{Ob}\,\mathfrak{C}$, $\text{Hom}_{\mathfrak{C}^{\text{op}}}(a, a')$ is by definition the class $\text{Hom}_{\mathfrak{C}}(a', a)$. So, it is "the same category", but with "reversed arrows".

We shall be especially interested in *abelian categories*. Let us consider the basic notions needed for their introduction.

Definition 1.3.

- $f \in \text{Hom}_{\mathfrak{C}}(a, a')$ is a *monomorphism* if it is left effaceable, i.e., for all $g_1, g_2 \in \text{Hom}_{\mathfrak{C}}(a'', a)$, the condition $f \circ g_1 = f \circ g_2$ implies $g_1 = g_2$;
- f is an *epimorphism* if it is right effaceable, i.e., for all $g_1, g_2 \in \text{Hom}_{\mathfrak{C}}(a', a'')$, the condition $g_1 \circ f = g_2 \circ f$ implies $g_1 = g_2$;
- it is an *isomorphism* if there exists $g \in \text{Hom}_{\mathfrak{C}}(a', a)$ such that $f \circ g = \text{id}_{a'}$ and $g \circ f = \text{id}_a$.

---

[a]Such a category is a *groupoid*, i.e., a small category in which every morphism is invertible.

If f is an isomorphism, it is both a monomorphism and an epimorphism. If $f : a \to a'$ is an isomorphism, the objects a and a' are said to be *isomorphic*.

Definition 1.4. A category $\mathfrak{A}$ is *additive* if

- all Hom classes are abelian groups, and the composition of morphisms is bilinear; that is, if a, a', a'' are objects in $\mathfrak{A}$, and f, $f' \in \mathrm{Hom}_{\mathfrak{A}}(a, a')$ and g, $g' \in \mathrm{Hom}_{\mathfrak{A}}(a', a'')$, then

$$(g + g') \circ (f + f') = g \circ f + g \circ f' + g' \circ f + g' \circ f';$$

- it has finite direct sums and direct products;
- it has a zero object.

Remark 1.5. In view of its very definition, every additive category is locally small.

Definition 1.6. An additive category $\mathfrak{A}$ is *abelian* if

- every morphism has kernel and cokernel[b];
- every monomorphism is the kernel of a morphism, and every epimorphism is the cokernel of a morphism.

The kernel and cokernel of a morphism f are denoted $\ker f$ and $\mathrm{coker}\, f$. If $f : a \to b$ is a monomorphism, the cokernel of f is also called the *quotient* of b by a, denoted b/a.

In an abelian category, a morphism is a monomorphism if and only if it has a left inverse: i.e., $f : a \to a'$ is a monomorphism if and only if there is a morphism $\ell : a' \to a$ such that $l \circ f = \mathrm{id}_a$. Analogously, a morphism in an abelian category is an epimorphism if and only if it has a right inverse.

Definition 1.7. If $f : a \to b$ is a morphism in an abelian category $\mathfrak{A}$, then the image of f, denoted $\mathrm{im}\, f$, is the kernel of the morphism $b \to \mathrm{coker}\, f$.

---

[b]Given a morphism $f : a \to b$, a kernel of f is an object k in $\mathfrak{A}$ and a morphism $i : k \to a$ such that

- $f \circ i = 0$;
- if $j : h \to a$ is a morphism in $\mathfrak{A}$ such that $f \circ j = 0$, then there is a unique morphism $g : h \to k$ such that $j = i \circ g$;

that is, every morphism into a which goes to zero in b factors uniquely through the kernel. The definition of cokernel is similar: it is an object c in $\mathfrak{A}$ with a morphism $p : b \to c$ such that $p \circ f = 0$, and if $q : b \to c'$ is a morphism such that $q \circ f = 0$, then q uniquely factors through c, i.e., there is a unique morphism $g : c \to c'$ such that $q = g \circ p$.

Example 1.8. The categories $\mathfrak{Ab}$ of abelian groups, R-**mod** of modules over a commutative ring R, $\mathfrak{Sh}_X$ of sheaves of abelian groups over a topological space X, $\mathcal{O}_X$-**mod** of modules over the sheaf of rings $\mathcal{O}_X$ of a ringed space $(X, \mathcal{O}_X)$, are all abelian categories (sheaves of abelian groups will form the object of Chapter 3, while ringed spaces and $\mathcal{O}_X$-modules will be introduced in Chapter 4).

Example 1.9. Let X be a topological space, and let $\mathbb{Z}_X$ be the locally constant sheaf on X associated with $\mathbb{Z}$ (i.e., the sheaf of locally constant $\mathbb{Z}$-valued functions). Then $(X, \mathbb{Z}_X)$ is a ringed space.

1.1.2. Functors and natural transformations

Loosely speaking, functors are "morphisms" between categories. For instance, taking the dual of a vector space defines a functor from the category of vector spaces on a fixed field to itself (one should also specify the action of the functor on the morphisms of the category, and in this case the functor takes a linear transformation to its dual).

Let $\mathfrak{A}$ and $\mathfrak{B}$ be categories. A (covariant) functor $F : \mathfrak{A} \to \mathfrak{B}$ consists of

- for each object $a \in \mathrm{Ob}\,\mathfrak{A}$, an object $F(a) \in \mathrm{Ob}\,\mathfrak{B}$;
- for each morphism $f \in \mathrm{Hom}_{\mathfrak{A}}(a, a')$, a morphism

$$F(f) \in \mathrm{Hom}_{\mathfrak{B}}(F(a), F(a')),$$

such that

- for all $a \in \mathrm{Ob}\,\mathfrak{A}$, $F(\mathrm{id}_a) = \mathrm{id}_{F(a)}$;
- for all $a, a', a'' \in \mathrm{Ob}\,\mathfrak{A}$ and $f \in \mathrm{Hom}_{\mathfrak{A}}(a, a')$ and $g \in \mathrm{Hom}_{\mathfrak{A}}(a', a'')$, one has

$$F(g \circ f) = F(g) \circ F(f). \tag{1.1}$$

Given a category $\mathfrak{C}$, there is a functor $\mathrm{id}_{\mathfrak{C}} : \mathfrak{C} \to \mathfrak{C}$, the *identity functor of* $\mathfrak{C}$, defined by the conditions $\mathrm{id}_{\mathfrak{C}}(a) = a$ for all $a \in \mathrm{Ob}\,\mathfrak{C}$, and $\mathrm{id}_{\mathfrak{C}}(f) = f$ for all $a, a' \in \mathrm{Ob}\,\mathfrak{C}$ and $f \in \mathrm{Hom}_{\mathfrak{C}}(a, a')$.

A contravariant functor $\mathfrak{A} \to \mathfrak{B}$ is a covariant functor $\mathfrak{A}^{\mathrm{op}} \to \mathfrak{B}$. So, given $a, a' \in \mathrm{Ob}\,\mathfrak{A}$, and $f \in \mathrm{Hom}_{\mathfrak{A}}(a, a')$, the morphism $F(f)$ is in $\mathrm{Hom}_{\mathfrak{B}}(F(a'), F(a))$, and instead of (1.1), one has

$$F(g \circ f) = F(f) \circ F(g). \tag{1.2}$$

Given categories $\mathfrak{A}$, $\mathfrak{B}$, and functors $F, G : \mathfrak{A} \to \mathfrak{B}$, a *natural transformation* $\eta : F \to G$ is a morphism $\eta_a \in \mathrm{Hom}_{\mathfrak{B}}(F(a), G(a))$ for every $a \in \mathrm{Ob}\,\mathfrak{A}$, such that for all $f \in \mathrm{Hom}_{\mathfrak{A}}(a, a')$, the diagram

$$
\begin{array}{ccc}
F(a) & \xrightarrow{\ F(f)\ } & F(a') \\
{\scriptstyle \eta_a} \downarrow & & \downarrow {\scriptstyle \eta_{a'}} \\
G(a) & \xrightarrow{\ G(f)\ } & G(a')
\end{array}
$$

commutes. In a way, a natural transformation from F to G is "a set of morphisms which make the functor F into the functor G". Natural transformations are also called *morphisms of functors*.

Definition 1.10. Two functors $F, G : \mathfrak{A} \to \mathfrak{B}$ are *naturally isomorphic* if there is a natural transformation $\eta : F \to G$ such that η_a is an isomorphism for all $a \in \mathrm{Ob}\,\mathfrak{A}$.

Definition 1.11. Two categories $\mathfrak{A}$ and $\mathfrak{B}$ are *isomorphic* if there are functors $F : \mathfrak{A} \to \mathfrak{B}$, $G : \mathfrak{B} \to \mathfrak{A}$ such that $G \circ F = \mathrm{id}_{\mathfrak{A}}$ and $F \circ G = \mathrm{id}_{\mathfrak{B}}$.

This notion may be too strong, and often a suppler way of "identifying" categories may be useful.

Definition 1.12. Two categories $\mathfrak{A}$ and $\mathfrak{B}$ are *equivalent* if there are functors $F : \mathfrak{A} \to \mathfrak{B}$, $G : \mathfrak{B} \to \mathfrak{A}$ and natural isomorphisms $\epsilon : F \circ G \to \mathrm{id}_{\mathfrak{B}}$ and $\eta : \mathrm{id}_{\mathfrak{A}} \to G \circ F$.

So, in the last condition, we ask that $\epsilon_b : F(G(b)) \to b$ is an isomorphism for all $b \in \mathrm{Ob}\,\mathfrak{B}$, and $\eta_a : a \to G(F(a))$ is an isomorphism for all $a \in \mathrm{Ob}\,\mathfrak{A}$.

By its very definition, for every pair of objects $a, a' \in \mathrm{Ob}\,\mathfrak{A}$, a functor $F : \mathfrak{A} \to \mathfrak{B}$ defines a map

$$\mathrm{Hom}_{\mathfrak{A}}(a, a') \to \mathrm{Hom}_{\mathfrak{B}}(F(a), F(a')). \tag{1.3}$$

The functor F is said to be

- *full* if the map (1.3) is surjective;
- *faithful* if the map (1.3) is injective;
- *fully faithful* if the map (1.3) is bijective;
- *essentially surjective* if every $b \in \mathrm{Ob}\,\mathfrak{B}$ is isomorphic to $F(a)$ for some $a \in \mathrm{Ob}\,\mathfrak{A}$.

Equivalences of categories can be characterized in the following way (for a proof see [43]).

Proposition 1.13. *A functor $F : \mathfrak{A} \to \mathfrak{B}$ defines an equivalence of categories if and only if it is fully faithful and essentially surjective.*

If $\mathfrak{A}$ and $\mathfrak{B}$ are categories, two functors $L : \mathfrak{A} \to \mathfrak{B}$ and $R : \mathfrak{B} \to \mathfrak{A}$ are said to be a pair of *adjoint functors* if for every pair of objects $a \in \mathrm{Ob}\,\mathfrak{A}$, $b \in \mathrm{Ob}\,\mathfrak{B}$, there is a natural isomorphism

$$\phi_{a,b} : \mathrm{Hom}_{\mathfrak{B}}(L(a), b) \to \mathrm{Hom}_{\mathfrak{A}}(a, R(b)).$$

Naturality means that for every morphism $g : b \to b'$ in $\mathfrak{B}$, and $f : a' \to a$ in $\mathfrak{A}$, the diagrams

$$
\begin{array}{ccc}
\mathrm{Hom}_{\mathfrak{B}}(L(a), b) & \xrightarrow{\phi_{a,b}} & \mathrm{Hom}_{\mathfrak{A}}(a, R(b)) \\
\downarrow & & \downarrow \\
\mathrm{Hom}_{\mathfrak{B}}(L(a), b') & \xrightarrow{\phi_{a,b'}} & \mathrm{Hom}_{\mathfrak{A}}(a, R(b'))
\end{array}
\tag{1.4}
$$

$$
\begin{array}{ccc}
\mathrm{Hom}_{\mathfrak{B}}(L(a), b) & \xrightarrow{\phi_{a,b}} & \mathrm{Hom}_{\mathfrak{A}}(a, R(b)) \\
\downarrow & & \downarrow \\
\mathrm{Hom}_{\mathfrak{B}}(L(a'), b) & \xrightarrow{\phi_{a',b}} & \mathrm{Hom}_{\mathfrak{A}}(a', R(b))
\end{array}
\tag{1.5}
$$

commute; here the vertical arrows are induced by the morphisms f and g. The functor L is said to be *left adjoint* to R and R *right adjoint* to L.

Example 1.14. If R is a commutative ring, any R-module M can be regarded as an abelian group, and any morphism of R-modules is in particular a morphism of abelian groups. This defines a "forgetful functor" $FF : R\text{-}\mathbf{mod} \to \mathfrak{Ab}$. If R has a unity, this functor has a right adjoint $H : \mathfrak{Ab} \to R\text{-}\mathbf{mod}$ defined as follows. Note that if G is an abelian group, and R a ring, the group $\mathrm{Hom}_{\mathbb{Z}}(R, G)^{\mathrm{c}}$ has an R-module structure given by

---

[c]As abelian groups can be seen as $\mathbb{Z}$-modules, we will write $\mathrm{Hom}_{\mathbb{Z}}$ instead of $\mathrm{Hom}_{\mathfrak{Ab}}$.

$(r\rho)(s) = \rho(rs)$ for all $r, s \in R$ and $\rho \in \mathrm{Hom}_{\mathbb{Z}}(R, G)$. Then set

$$H(G) = \mathrm{Hom}_{\mathbb{Z}}(R, G).$$

The required isomorphism

$$\phi : \mathrm{Hom}_{\mathbb{Z}}(FF(M), G) \to \mathrm{Hom}_R(M, H(G))$$

is given by

$$\phi(\eta)(m)(r) = \eta(rm)$$

for every morphism of abelian groups $\eta : M \to G$ and every $m \in M, r \in R$. The inverse isomorphism

$$\psi : \mathrm{Hom}_R(M, H(G)) \to \mathrm{Hom}_{\mathbb{Z}}(FF(M), G)$$

is

$$\psi(\gamma)(m) = \gamma(m)(1)$$

for every morphism of R-modules $\gamma : M \to H(G)$ and every $m \in M$.

1.2. Elements of Homological Algebra

Homological algebra studies the notions of homology and cohomology in a general, abstract algebraic setting. It originates from algebraic topology and commutative algebra.

The aim of this section is to provide the reader with the basic notions in homological algebra. Standard references are [30, 41, 64].

1.2.1. Exact sequences

Let $\mathfrak{A}$ be an abelian category, and let a, a', a'' be objects in $\mathfrak{A}$. Moreover, let $i : a' \to a$, $p : a \to a''$ be morphisms. If $p \circ i = 0$, by the definition of kernel (in particular, its universality), there is a natural morphism $\mathrm{im}\, i \to \ker p$. We say that the pair (i, p) forms an exact sequence, and write

$$0 \to a' \xrightarrow{i} a \xrightarrow{p} a'' \to 0,$$

if

- i is a monomorphism,
- p is an epimorphism,
- $p \circ i = 0$ and the natural morphism $\mathrm{im}\, i \to \ker p$ is an isomorphism.

Exercise 1.15. If R is a commutative ring, and $i : M' \to M$, $p : M \to M''$ are morphisms in the category of R-modules R-**mod**, prove that

$$0 \to M' \xrightarrow{i} M \xrightarrow{p} M'' \to 0$$

is an exact sequence if and only if i is injective, p is surjective, and $\ker p = \operatorname{im} i$ as submodules of M.

Exercise 1.16. Set $R = \mathbb{Z}$, the ring of integers (recall that $\mathbb{Z}$-modules are just abelian groups), and consider the sequence

$$0 \to \mathbb{Z} \xrightarrow{j} \mathbb{C} \xrightarrow{\exp} \mathbb{C}^* \to 1,$$

where j is the inclusion of the integers into the complex numbers $\mathbb{C}$, while $\mathbb{C}^* = \mathbb{C} - \{0\}$ is the multiplicative group of nonzero complex numbers. The morphism exp is defined as $\exp(z) = e^{2\pi i z}$. The reader may check that this sequence is exact.

A morphism of exact sequences is a commutative diagram

$$
\begin{array}{ccccccccc}
0 & \longrightarrow & a' & \longrightarrow & a & \longrightarrow & a'' & \longrightarrow & 0 \\
 & & \downarrow & & \downarrow & & \downarrow & & \\
0 & \longrightarrow & b' & \longrightarrow & b & \longrightarrow & b'' & \longrightarrow & 0
\end{array}
$$

of morphisms in $\mathfrak{A}$ whose rows are exact.

1.2.2. Differential complexes

Differential complexes, also called chain complexes, are the basic objects of study of homological algebra.

Let $\mathfrak{A}$ be an abelian category.

Definition 1.17. A differential on an object a of $\mathfrak{A}$ is a morphism $d : a \to a$ in $\mathfrak{A}$ such that $d^2 \equiv d \circ d = 0$. The pair (a, d) is called a differential object in $\mathfrak{A}$.

We shall denote $Z(a, d) = \ker d$ and $B(a, d) = \operatorname{im} d$. Since $d^2 = 0$, there is a monomorphism $B(a, d) \to Z(a, d)$. The quotient

$$H(a, d) = Z(a, d)/B(a, d)$$

is called the *cohomology object* of the differential object (a, d).

Remark 1.18. If $\mathfrak{A} = R$-**mod** (or more generally, whenever the objects of $\mathfrak{A}$ are sets)[d] the elements of a, $Z(a,d)$ and $B(a,d)$ are called *cochains*, *cocycles* and *coboundaries* of (a,d), respectively. The cohomology object $H(a,d)$ is an R-module.

We shall often write $Z(a)$, $B(a)$ and $H(a)$, omitting the differential d, when there is no risk of confusion.

Let (a,d) and (a',d') be differential objects in $\mathfrak{A}$.

Definition 1.19. A morphism of differential objects is a morphism $f :$ $a \to a'$ in $\mathfrak{A}$ which commutes with the differentials, $f \circ d = d' \circ f$.

The morphism f maps $Z(a,d)$ to $Z(a',d')$ and $B(a,d)$ to $B(a',d')$, thus inducing a morphism $H(f) : H(a) \to H(a')$.

The following proposition is a basic construction in homological algebra.

Proposition 1.20. *Let* $0 \to a' \overset{i}{\to} a \overset{p}{\to} a'' \to 0$ *be an exact sequence of differential objects in* $\mathfrak{A}$. *Then there exist a morphism* $\delta : H(a'') \to H(a')$ (*called connecting morphism*) *and an exact triangle of cohomology*[e]

$$
\begin{array}{ccc}
H(a) & \xrightarrow{\;H(p)\;} & H(a'') \\
{\scriptstyle H(i)}\big\uparrow & \swarrow {\scriptstyle \delta} & \\
H(a') & &
\end{array}
$$

Proof. For simplicity we sketch a proof when $\mathfrak{A}$ is a concrete category, for instance R-**mod**; there is no conceptual difference from the general case. The construction of δ is as follows: let $\xi'' \in H(a'')$ and let m'' be a cocycle whose class is ξ''. If m is an element of a such that $p(m) = m''$, we have $p(dm) = d''m'' = 0$ and then $dm = i(m')$ for some $m' \in a'$ which is a cocycle. Now, the cocycle m' defines a cohomology class $\delta(\xi'')$ in $H(a')$, which is independent of the choices we have made, thus defining a morphism

---

[d]More precisely, one does not only need that the objects are sets, but also that the morphisms in the category are morphisms of sets, satisfying suitable additional requirements. In other terms, there is a fully faithful functor from the category in question to the category of sets. Such categories are called *concrete*. If a category can be equipped with such a functor it is said to be *concretizable*. Not all categories are concretizable; for instance, the category whose objects are topological spaces, and morphisms are homotopy classes of continuous morphisms, is not concretizable [18].

[e]One says that a triangle of morphisms is exact when the kernel of every arrow is isomorphic to the image of the previous one.

$\delta : H(a'') \to H(a')$. One proves by direct computation that the triangle is exact. A proof of these facts is given for instance in [30, Section IV.2]. $\square$

Actually, what we have been using is the Snake Lemma, which is proved in Appendix A.2 for a general abelian category.

The above results can be translated to the setting of complexes in $\mathfrak{A}$.

Definition 1.21. A complex in $\mathfrak{A}$ is a family $a^\bullet = \{a^n\}_{n \in \mathbb{Z}}$ of objects in $\mathfrak{A}$ with morphisms $d_n : a^n \to a^{n+1}$ such that $d_{n+1} \circ d_n = 0$ for every $n \in \mathbb{Z}$.

The morphisms d_n are the *differentials* of the complex. We shall usually write a complex in $\mathfrak{A}$ in the more pictorial form

$$\cdots \xrightarrow{d_{n-2}} a^{n-1} \xrightarrow{d_{n-1}} a^n \xrightarrow{d_n} a^{n+1} \xrightarrow{d_{n+1}} \cdots .$$

For every n one defines the objects $Z^n(a^\bullet) = \ker d_n$ (the nth cocycle object) and $B^n(a^\bullet) = \operatorname{im} d_{n-1}$ (the nth coboundary object), and also the nth cohomology object $H^n(a^\bullet) = Z^n(a^\bullet)/B^n(a^\bullet)$; this makes sense as $d_n \circ d_{n-1} = 0$. When $\mathfrak{A} = R\text{-mod}$, the cohomology objects are usually called *cohomology modules (or groups)*.

Definition 1.22. A morphism of complexes in $\mathfrak{A}$, $f : a^\bullet \to b^\bullet$, is a collection of morphisms $\{f_n : a^n \to b^n | n \in \mathbb{Z}\}$, such that the following diagram commutes:

$$
\begin{array}{ccc}
a^n & \xrightarrow{f_n} & b^n \\
d \downarrow & & \downarrow d \\
a^{n+1} & \xrightarrow{f_{n+1}} & b^{n+1}
\end{array}
$$

Given morphisms of complexes $i : a^\bullet \to b^\bullet$ and $p : b^\bullet \to c^\bullet$, we say that they form an exact sequence of complexes, and write

$$0 \to a^\bullet \xrightarrow{i} b^\bullet \xrightarrow{p} c^\bullet \to 0,$$

if for every n the sequence

$$0 \to a^n \xrightarrow{i_n} b^n \xrightarrow{p_n} c^n \to 0$$

is exact.

For complexes, Proposition 1.20 takes the following form.

Proposition 1.23. *Let* $0 \to a^\bullet \xrightarrow{i} b^\bullet \xrightarrow{p} c^\bullet \to 0$ *be an exact sequence of complexes in* $\mathfrak{A}$. *Then there exist connecting morphisms* $\delta_n : H^n(c^\bullet) \to H^{n+1}(a^\bullet)$ *and a long exact sequence of cohomology*

$$\cdots \xrightarrow{\delta_{n-1}} H^n(a^\bullet) \xrightarrow{H(i)} H^n(b^\bullet) \xrightarrow{H(p)} H^n(c^\bullet)$$

$$\xrightarrow{\delta_n} H^{n+1}(a^\bullet) \xrightarrow{H(i)} H^{n+1}(b^\bullet) \xrightarrow{H(p)} H^{n+1}(c^\bullet) \xrightarrow{\delta_{n+1}} \cdots.$$

Proof. The connecting morphism $\delta : H^\bullet(c^\bullet) \to H^\bullet(a^\bullet)$ defined in Proposition 1.20 splits into morphisms $\delta_n : H^n(c^\bullet) \to H^{n+1}(a^\bullet)$ (indeed the connecting morphism increases the degree by one) and the long exact sequence of the statement is obtained by developing the exact triangle of cohomology. $\square$

Exercise 1.24. A morphism of exact sequences of complexes is a commutative diagram

$$
\begin{array}{ccccccccc}
0 & \longrightarrow & a^\bullet & \longrightarrow & b^\bullet & \longrightarrow & c^\bullet & \longrightarrow & 0 \\
& & \downarrow{\scriptstyle f} & & \downarrow{\scriptstyle g} & & \downarrow{\scriptstyle h} & & \\
0 & \longrightarrow & a'^\bullet & \longrightarrow & b'^\bullet & \longrightarrow & c'^\bullet & \longrightarrow & 0
\end{array}
$$

where the rows are exact sequences of complexes, and the vertical arrows are morphisms of complexes. Let $\{\delta_i\}$, $\{\delta_i'\}$ be the connecting morphisms associated with the two exact sequences. Show that for every $i \geq 0$ the square

$$
\begin{array}{ccc}
H^i(c^\bullet) & \xrightarrow{\delta_i} & H^{i+1}(a^\bullet) \\
{\scriptstyle H^i(h)}\downarrow & & \downarrow{\scriptstyle H^{i+1}(f)} \\
H^i(c'^\bullet) & \xrightarrow{\delta_i'} & H^{i+1}(a'^\bullet)
\end{array}
$$

commutes.

1.2.3. Homotopies

Different (i.e., nonisomorphic) complexes may nevertheless have isomorphic cohomologies. A sufficient condition for this to hold is that the two complexes are *homotopic*. While this condition is not necessary, in practice the (by far) commonest way to prove the isomorphism between two cohomologies is to exhibit a homotopy between the corresponding complexes.

Definition 1.25. Given two complexes in $\mathfrak{A}$, $(a^\bullet, d)$ and $(b^\bullet, d')$, and two morphisms of complexes, $f, g : a^\bullet \to b^\bullet$, a *homotopy* between f and g is a morphism $K : a^\bullet \to b^{\bullet-1}$ (i.e., for every k, a morphism $K : a^k \to b^{k-1}$) such that $d' \circ K + K \circ d = f - g$.

The situation is depicted in the following commutative diagram:

$$
\begin{array}{ccccccc}
\cdots \longrightarrow & a^{k-1} & \xrightarrow{\ d\ } & a^k & \xrightarrow{\ d\ } & a^{k+1} & \longrightarrow \cdots \\
& \Big\Downarrow & K \swarrow f \Big\downarrow g \ K \swarrow & & & \Big\Downarrow & \\
\cdots \longrightarrow & b^{k-1} & \xrightarrow{\ d'\ } & b^k & \xrightarrow{\ d'\ } & b^{k+1} & \longrightarrow \cdots
\end{array}
$$

Proposition 1.26. *If there is a homotopy between f and g, then $H(f) = H(g)$, namely, homotopic morphisms induce the same morphism in cohomology.*

Proof. If $\mathfrak{A} = R\text{-mod}$, the proof can be written as follows. Let $\xi = [m] \in H^k(a^\bullet, d)$, with $m \in Z^k(a, d)$. Then

$$H(f)(\xi) = [f(m)] = [g(m)] + [d'K(m)] + [K(dm)] = [g(m)] = H(g)(\xi)$$

since $dm = 0$ and $[d'K(m)] = 0$.

In the general case, we note that the morphisms $H(f)$ and $H(g)$ fit into the diagram

$$
\begin{array}{ccc}
Z^k(a, d) & \overset{f}{\underset{g}{\rightrightarrows}} & Z^k(b, d') \\
\downarrow & & \downarrow \\
H^k(a, d) & \overset{H^k(f)}{\underset{H^k(g)}{\rightrightarrows}} & H^k(b, d') \\
\downarrow & & \downarrow \\
0 & & 0
\end{array}
$$

Since d' takes values in $B^k(b, d')$, we have $H(d' \circ K) = 0$. Moreover, $d = 0$ on $Z^k(a, d)$, so that $H(K \circ d) = 0$. Hence, $H(f) = H(g)$. $\qquad\square$

Definition 1.27. Two complexes $(a^\bullet, d)$ and $(b^\bullet, d')$ in $\mathfrak{A}$ are said to be *homotopically equivalent* (or *homotopic*) if there exist morphisms $f : a^\bullet \to b^\bullet$, $g : b^\bullet \to a^\bullet$, such that:

$f \circ g : b^\bullet \to b^\bullet$ is homotopic to the identity map id_b;
$g \circ f : a^\bullet \to a^\bullet$ is homotopic to the identity map id_a.

Corollary 1.28. *Two homotopic complexes have isomorphic cohomologies.*

Proof. According to Definition 1.27, one has

$$H(f) \circ H(g) = H(f \circ g) = H(\mathrm{id}_b) = \mathrm{id}_{H(b)}$$
$$H(g) \circ H(f) = H(g \circ f) = H(\mathrm{id}_a) = \mathrm{id}_{H(a)}$$

so that both $H(f)$ and $H(g)$ are isomorphism. □

Definition 1.29. A null homotopy of a complex $(a^\bullet, d)$ in $\mathfrak{A}$ is a homotopy between the identity morphism of $a^\bullet$ and the zero morphism; more explicitly, it is a morphism $K : a^\bullet \to a^{\bullet-1}$ such that $d \circ K + K \circ d = \mathrm{id}_a$.

Proposition 1.30. *If a complex $(a^\bullet, d)$ in $\mathfrak{A}$ admits a null homotopy, then it is exact (i.e., all its cohomology groups vanish; one also says that the complex is acyclic).*

Proof. Apply Corollary 1.28, or note that

$$\mathrm{id}_{H(a)} = H(\mathrm{id}_a) = 0$$

so that $H(a, d) = 0$. □

Remark 1.31. More generally, one can state that if a null homotopy $K : a^k \to a^{k-1}$ exists for $k \geq k_0$, then $H^k(a, d) = 0$ for $k \geq k_0$. In the case of complexes bounded below zero (i.e., $a^n = 0$ for $n < 0$), often a homotopy is defined only for $k \geq 1$, and it may happen that $H^0(a, d) \neq 0$.

Remark 1.32. One might as well define a homotopy by requiring $d' \circ K - K \circ d = f - g$; the reader may easily check that this change of sign is immaterial.

1.2.4. Left and right exact functors

Exact sequences are an important structure associated with abelian categories. It is therefore quite natural to investigate the behaviour of functors with respect to exact sequences. In a sense, this will be at the heart of the notion of derived functor.

A covariant functor $F : \mathfrak{A} \to \mathfrak{B}$ between abelian categories is said to be *additive* if for any two objects $a, a' \in \mathrm{Ob}\,\mathfrak{A}$, the induced map $\mathrm{Hom}(a, a') \to \mathrm{Hom}(F(a), F(a'))$ is a homomorphism of abelian groups. It is not difficult to show that this condition implies that

$$F(a \oplus a') \simeq F(a) \oplus F(a');$$

for a detailed discussion of this issue see, for example, [49, Lemma 12.7].

Definition 1.33. A covariant functor $F : \mathfrak{A} \to \mathfrak{B}$ is said to be *left exact* if for any short exact sequence

$$0 \to a' \to a \to a'' \to 0$$

in $\mathfrak{A}$, the sequence

$$0 \to F(a') \to F(a) \to F(a'')$$

is exact in $\mathfrak{B}$.

$F : \mathfrak{A} \to \mathfrak{B}$ is said to be *right exact* if for any short exact sequence

$$0 \to a' \to a \to a'' \to 0$$

in $\mathfrak{A}$, the sequence

$$F(a') \to F(a) \to F(a'') \to 0$$

in $\mathfrak{B}$ is exact.

One says F is *exact* if it is both left exact and right exact, i.e., the sequence

$$0 \to F(a') \to F(a) \to F(a'') \to 0$$

is exact.

In the case of contravariant functors, the definitions are similar, changing the directions of the arrows; in other terms, a contravariant functor is left or right exact if it is so as a covariant functor from the opposite category.

We make now a number of examples that will be very useful later on.

Example 1.34. If $\mathfrak{A}$ is an abelian category, the functor

$$\mathrm{Hom}_{\mathfrak{A}}(b, -) : \mathfrak{A} \to \mathfrak{A}\mathfrak{b}$$

$$a \mapsto \mathrm{Hom}_{\mathfrak{A}}(b, a),$$

where b is a fixed object in $\mathfrak{A}$, is left exact. We prove this claim. Consider an exact sequence

$$0 \to a' \xrightarrow{i} a \xrightarrow{p} a'' \to 0. \tag{1.6}$$

We want to prove that

$$0 \to \operatorname{Hom}_{\mathfrak{A}}(b, a') \to \operatorname{Hom}_{\mathfrak{A}}(b, a) \to \operatorname{Hom}_{\mathfrak{A}}(b, a'')$$

is exact. For brevity, we denote $\operatorname{Hom}_{\mathfrak{A}}(b, -)$, for $b \in \operatorname{Ob} \mathfrak{A}$, as F. So, we need to prove that

- $F(i) : F(a') \to F(a)$ is a monomorphism (i.e., an injective morphism of abelian groups);
- $\ker F(p) = \operatorname{im} F(i)$.

The first condition follows from the fact that i has a left inverse g, so that $g \circ i = \operatorname{id}_{a'}$, and $F(g) \circ F(i) = \operatorname{id}_{F(a')}$, i.e., $F(i)$ has a left inverse. For the second condition, first we note that $p \circ i = 0$ implies $F(p) \circ F(i) = 0$, so that $\operatorname{im} F(i) \subset \ker F(p)$. On the other hand, if $f \in \ker F(p)$, i.e., $p \circ f = 0$, then f factors through $\ker p = \operatorname{im} i$, so that $f \in \operatorname{im} F(i)$.

Example 1.35. In particular, we may consider:

- $\operatorname{Hom}_{R\text{-}\mathbf{mod}}(M, -) : R\text{-}\mathbf{mod} \to R\text{-}\mathbf{mod}$, where $R\text{-}\mathbf{mod}$ is the category of modules over a commutative ring R. Actually, we shall denote $\operatorname{Hom}_{R\text{-}\mathbf{mod}}$ by Hom_R.
- $\operatorname{Hom}_{\mathfrak{Sh}_X}(\mathcal{F}, -) : \mathfrak{Sh}_X \to \mathfrak{Ab}$, where $\mathfrak{Sh}_X$ is the category of sheaves of abelian groups on a topological space X, that we shall introduce in Section 3.1.
- $\operatorname{Hom}_{\mathcal{O}_X}(\mathcal{M}, -) : \mathcal{O}_X\text{-}\mathbf{mod} \to \mathfrak{Ab}$, where $\mathcal{O}_X\text{-}\mathbf{mod}$ is the category of sheaves of $\mathcal{O}_X$-modules with $(X, \mathcal{O}_X)$ a ringed space, that we shall introduce in Section 4.2.

All these functors are additive and left exact.

Example 1.36. Analogously, the contravariant functor

$$\operatorname{Hom}_{\mathfrak{A}}(-, b) : \mathfrak{A} \to \mathfrak{Ab}$$

$$a \mapsto \operatorname{Hom}_{\mathfrak{A}}(a, b),$$

thought of as a covariant functor $\mathfrak{A}^{\mathrm{op}} \to \mathfrak{Ab}$, is left exact, that is, an exact sequence as in (1.6) produces the exact sequence

$$0 \to \operatorname{Hom}_{\mathfrak{A}}(a'', b) \to \operatorname{Hom}_{\mathfrak{A}}(a, b) \to \operatorname{Hom}_{\mathfrak{A}}(a', b).$$

The proof of this fact is completely analogous.

1.2.5. Exact sequences of functors

Let F, G and H be three functors from a category $\mathfrak{A}$ to an abelian category $\mathfrak{B}$, and let $F \xrightarrow{i} G$, $G \xrightarrow{p} H$ be natural transformations. One says that the sequence

$$0 \to F \xrightarrow{i} G \xrightarrow{p} H \to 0$$

is exact if for every object $a \in \mathrm{Ob}\,\mathfrak{A}$ the sequence of morphisms in $\mathfrak{B}$

$$0 \to F(a) \xrightarrow{i_a} G(a) \xrightarrow{p_a} H(a) \to 0$$

is exact.

Example 1.37. Let $\mathfrak{A} = \mathfrak{Op}(X)$ be the category of open subsets of a topological space X, with inclusions as morphisms, and $\mathfrak{B} = \mathfrak{Ab}$ the category of abelian groups. A *presheaf of abelian groups* is a functor $\mathcal{P} : \mathfrak{Op}(X)^{\mathrm{op}} \to \mathfrak{Ab}$. For every $U, V \in \mathfrak{Op}(X)$ with a morphism $i : U \to V$ (i.e., for every pair of nested open subsets in X, that is, $U \subset V \subset X$), there is a morphism of abelian groups

$$\mathcal{P}(i) : \mathcal{P}(V) \to \mathcal{P}(U),$$

called *restriction morphism*. According to the definition of functor, if W is an open subset of U, with inclusion $j : W \to U$, one must have

$$\mathcal{P}(i \circ j) = \mathcal{P}(j) \circ \mathcal{P}(i) \tag{1.7}$$

(cf. equation (1.2)), and

$$\mathcal{P}(\mathrm{id}_U) = \mathrm{id}_{\mathcal{P}(U)}.$$

The restriction morphisms are usually denoted $\mathcal{P}(i) = \rho_{V,U}$, so that the condition (1.7) amounts to the commutativity of the diagram

$$
\begin{array}{ccc}
\mathcal{P}(V) & \xrightarrow{\rho_{V,U}} & \mathcal{P}(U) \\
& \rho_{V,W} \searrow & \downarrow \rho_{U,W} \\
& & \mathcal{P}(W)
\end{array}
$$

If $\mathcal{P}$, $\mathcal{Q}$ are presheaves of abelian groups, a natural transformation f between the functors $\mathcal{P}$ and $\mathcal{Q}$ is called a *morphism of presheaves of abelian groups*; it consists of a collection $\{f_U\}_{U \in \mathrm{Ob}\,\mathfrak{Op}(X)}$ (i.e., a morphism for every open set U in X), where $f_U : \mathcal{P}(U) \to \mathcal{Q}(U)$. If $U \subset V$ are open sets in X, with inclusion $i : U \to V$, one must have

$$\mathcal{Q}(i) \circ f_V = f_U \circ \mathcal{P}(i),$$

i.e., denoting respectively by ρ and σ the restriction morphisms of $\mathcal{P}$ and $\mathcal{Q}$, the diagram

$$
\begin{array}{ccc}
\mathcal{P}(V) & \xrightarrow{f_V} & \mathcal{Q}(V) \\
{\scriptstyle \rho_{V,U}}\downarrow & & \downarrow{\scriptstyle \sigma_{V,U}} \\
\mathcal{P}(U) & \xrightarrow{f_U} & \mathcal{Q}(U)
\end{array}
$$

commutes.

Now, if we have presheaves of abelian groups $\mathcal{P}'$, $\mathcal{P}$, $\mathcal{P}''$ on X, with morphisms $i : \mathcal{P}' \to \mathcal{P}$, $p : \mathcal{P} \to \mathcal{P}''$, we say that they form an *exact sequence* of presheaves if

$$0 \to \mathcal{P}' \xrightarrow{i} \mathcal{P} \xrightarrow{p} \mathcal{P} \to 0$$

is an exact sequence of functors. This amounts to saying that for every open subset $U \subset X$ the sequence

$$0 \to \mathcal{P}'(U) \xrightarrow{i_U} \mathcal{P}(U) \xrightarrow{p_U} \mathcal{P}''(U) \to 0$$

is exact.

1.3. Additional Exercises

1. Prove that for all R-modules M and N, where R is a commutative ring, the group $\operatorname{Hom}_R(M, N)$ has an R-module structure given by $(r\rho)(m) = r\rho(m)$, for all $r \in R$, $m \in M$ and $\rho \in \operatorname{Hom}_R(M, N)$.

2. A nonzero element x in an abelian group G is said to be *torsion* if $nx = 0$ for some nonzero natural number n. G is said to be *torsion-free* if it has no torsion elements.

   (a) Define the category of torsion-free abelian groups.
   (b) Prove that this category is additive but not abelian.

3. (a) Define the category of finitely generated abelian groups.
   (b) Prove that it is abelian.

4. Show that the category of cyclic groups is not abelian.

5. A *monoid* is a set M with a binary operation $M \times M \to M$ which is associative and has an identity element. Let $\mathfrak{Mon}$ be the category of monoids, with the obvious morphisms. Show that in $\mathfrak{Mon}$ the natural morphism $\mathbb{N} \to \mathbb{Z}$ is an epimorphism.

6. (a) Prove that the category of commutative rings with unity is not abelian.
   (b) Show that in this category the unique morphism $\mathbb{Z} \to \mathbb{Q}$ is an epimorphism.

7. Check if the following forgetful functors are full, faithful or essentially surjective:

   (a) $\mathfrak{Gr} \to \mathfrak{Set}$, where $\mathfrak{Gr}$ is the category of groups and $\mathfrak{Set}$ the category of sets.
   (b) $\mathfrak{Top}_* \to \mathfrak{Top}$, where $\mathfrak{Top}_*$ and $\mathfrak{Top}$ are the categories of pointed topological spaces and topological spaces, respectively. (The objects of $\mathfrak{Top}_*$ are pairs (X, x), where X is a topological space and $x \in X$, and a morphism $f : (X, x) \to (Y, y)$ is a continuous map $f : X \to Y$ such that $f(x) = y$.)
   (c) $\mathfrak{Rng} \to \mathfrak{Gr}$, where $\mathfrak{Rng}$ is the category of rings.

8. An object i in a category $\mathfrak{C}$ is *initial* if for every object a there is exactly one morphism $i \to a$. An object t is *terminal* if for every object a there is exactly one morphism $a \to t$. Prove the following statements:

   (a) If $F : \mathfrak{C} \to \mathfrak{C}'$ is a full and faithful functor, and $F(a)$ is initial (terminal), then a initial (terminal).
   (b) Any two initial (terminal) objects in a category are isomorphic.

9. For each one of the following categories, determine if they admit initial and terminal objects, and characterize them:

   - $\mathfrak{Set}$ (sets);
   - $\mathfrak{Gr}$ (groups);
   - $\mathfrak{Top}$ (topological spaces);
   - $\mathfrak{Top}_*$ (pointed topological spaces);
   - $\mathfrak{Rng}$ (rings);
   - $\mathfrak{Rng}^1$ (rings with unity);
   - $\mathfrak{Fld}$ (fields);
   - R-**mod** (modules over a ring R).

10. A partially order set $(P, \leq)$ can be regarded as a category, whose objects are the elements of P; one says that there is a (unique) morphism $x \to y$ if and only if $x \leq y$.

    (a) Discuss under which conditions this category has an initial or terminal object.

(b) Given two preorders P and Q regarded as categories, characterize the functors $P \to Q$. Show that there exists a (unique) natural transformation between two such functors F, G if and only if $F(x) \leq G(x)$ for every $x \in P$.

11. Find an example of a category with no initial object, which has a subcategory having an initial object.

12. A monoid can be regarded as a category with one object, and morphisms given by the elements of the monoid; the composition of morphisms is given by the product of the monoid.

(a) Given two monoids M and N regarded as categories, describe the functors $F : M \to N$.

(b) Given two functors $F, G : M \to N$, characterize the natural transformations $F \to G$.

13. Prove Proposition 1.13.

14. Characterize the vertical arrows in the diagrams (1.4) and (1.5).

15. Let $\mathfrak{Rng}^c$ be the category of commutative rings. Consider $R, S \in \mathrm{Ob}\,\mathfrak{Rng}^c$ and $\rho \in \mathrm{Hom}_{\mathfrak{Rng}^c}(R, S)$.

(a) Prove that the assignment

$$F_R : S\text{-}\mathbf{mod} \to R\text{-}\mathbf{mod}$$

$$N \mapsto F_R(N),$$

which sends every S-module N to the R-module $F_R(N)$, which is N with the multiplication by elements of R defined by $r \cdot n = \rho(r)\, n$, is a functor.

(b) Prove that the functor $S \otimes_R - : R\text{-}\mathbf{mod} \to S\text{-}\mathbf{mod}$ is left adjoint to F_R.

16. Let $f_n : \mathbb{Z} \to \mathbb{Z}$ be the morphism given by the multiplication by a natural number n, $n \geq 2$. Prove that the sequence

$$0 \to \mathbb{Z} \xrightarrow{f_n} \mathbb{Z} \to \mathbb{Z}_n \to 0$$

is exact.

17. Prove that the category of complexes in an abelian category $\mathfrak{A}$ is abelian.

18. Consider the complex of vector spaces over a field $\Bbbk$

$$\frac{\Bbbk[x]}{\Bbbk} \to \Bbbk[x, y] \to \Bbbk,$$

where the first arrow is the obvious inclusion morphism and the second is the evaluation at zero. Compute its cohomology.

19. Let V be an n-dimensional vector space over a field $\Bbbk$, and fix a nonzero $v \in V$. The Koszul complex of the pair (V, v) is

$$0 \to \Lambda^n V^* \to \Lambda^{n-1} V^* \to \cdots \to V^* \to \Bbbk \to 0,$$

where all arrows are the inner product by v. Show that it is exact.

20. Given a field $\Bbbk$, let $R = \Bbbk[x, y, z]$. Let M be a free module over R of rank two. Show that there exists an exact sequence

$$0 \to \Lambda^2 M \to M \to R \to \Bbbk[z] \to 0.$$

This is the *free Koszul resolution* of the quotient ring $\Bbbk[z] = R/(x, y)$ as an R-module.

Hint: think of M as the free module generated by the symbols x and y, and take inspiration from Exercise 19.

21. Let $(C^i, d)_{i \in \mathbb{Z}}$ be a complex in $\mathfrak{Ab}$, where $C^i = \mathbb{Z}_9$ for all i, and the differential $d : C^\bullet \to C^{\bullet+1}$ is the multiplication by 3. Let us consider also the complex $(D^i, 0)_{i \in \mathbb{Z}}$, where again $D^i = \mathbb{Z}_9$.

(a) Compute the cohomology of the two complexes.

(b) Prove that $f : C^\bullet \to D^\bullet$ given by the multiplication by 3 is a morphism of complexes.

(c) Prove that id $: C^\bullet \to D^{\bullet-1}$ is a homotopy between f and the zero morphism.

Chapter 2

Derived Functors

Many interesting functors between abelian categories are left or right exact, but fail to be exact. Derived functors measure how far a left or right exact functor is from being exact. The notion of derived functor actually encompasses many cohomological theories and provides a powerful toolset for studying them. Important examples of derived functors are the Ext and Tor groups, sheaf cohomology groups, higher direct images, and more. Later in Chapter 5 we shall introduce a generalization of derived functors, the *hyperderived functors*.

2.1. Injective Objects

Derived functors are defined by means of a special class of objects, the *injective objects*.

Let $\mathfrak{A}$ be an abelian category. As we mentioned in Section 1.2.4, any object e in $\mathfrak{A}$ defines a left exact contravariant functor

$$\operatorname{Hom}(-, e) : \mathfrak{A} \to \mathfrak{Ab}$$

$$a \mapsto \operatorname{Hom}(a, e).$$

Definition 2.1. We say that an exact sequence in $\mathfrak{A}$

$$0 \to a' \xrightarrow{i} a \xrightarrow{p} a'' \to 0 \tag{2.1}$$

splits if there is a morphism $f : a \to a'$ such that $f \circ i = \operatorname{id}_{a'}$.

The morphism f is called a *retraction* of i.

Proposition 2.2.

(1) *If the exact sequence (2.1) splits, then*

$$a \simeq a' \oplus a''.$$

(2) *The exact sequence (2.1) splits if and only if there is a morphism*

$$s : a'' \to a$$

such that $p \circ s = \mathrm{id}_{a''}$.

The morphism s is called a *section* of p.

Proof. (1) The morphism $g = (f, p) : a \to a' \oplus a''$ is an isomorphism.

(2) Given the morphism s, one has $p \circ (\mathrm{id}_a - s \circ p) = 0$, so that by the definition of kernel, there is a morphism $f : a \to a'$ such that

$$i \circ f = \mathrm{id}_a - s \circ p.$$

Composing with i from the right we get $i \circ f \circ i = i$, and since i is left effaceable, $f \circ i = \mathrm{id}_{a'}$.

Conversely, if f is given, we have $(\mathrm{id}_a - i \circ f) \circ i = 0$, so that by the definition of cokernel there is morphism $s : a'' \to a$ such that

$$s \circ p = \mathrm{id}_a - i \circ f.$$

Composing on the left with p we get $p \circ s \circ p = p$, and as p is right effaceable, we obtain $p \circ s = \mathrm{id}_{a''}$.[a] $\qquad\square$

Exercise 2.3. One can use the notion of split exact sequence to show that left and right exact functors are additive.

---

[a]When $\mathfrak{A} = R\text{-}\mathbf{mod}$, this proof can be written as follows. Given the morphism s and $x \in a$, then $p(s(p(x)) - x) = 0$, so that $s(p(x)) - x = i(x')$ for a unique $x' \in a'$. Then set $f(x) = x'$.

Conversely, if f is given, and $x'' \in a''$, take a counterimage x of x'' in a, and set

$$s(x'') = x - i(f(x)).$$

This is well defined as, if y is another counterimage, then $y = x + i(y')$ for an element $y' \in a'$, so that

$$y - i(f(y)) = x + i(y') - i(f(x)) - i(f(i(y'))) = x - i(f(x)).$$

Moreover, $p(s(x'')) = x''$.

Proposition 2.4. $\mathrm{Hom}(-,e)$ *is an exact functor if and only if*

(1) *given a monomorphism* $j : a' \to a$ *and a morphism* $f : a' \to e$, *there exists* $g : a \to e$ *such that*

$$0 \longrightarrow a' \xrightarrow{\ j\ } a \qquad\qquad (2.2)$$

$$\downarrow f \quad \swarrow g$$

$$e$$

commutes[b]; *or*

(2) *every exact sequence*

$$0 \to e \xrightarrow{i} a \xrightarrow{p} a'' \to 0 \qquad\qquad (2.3)$$

splits.

Proof. (1) We already know that $\mathrm{Hom}(-,e)$ is left exact, so what we need to show is that, given an exact sequence

$$0 \to a' \xrightarrow{j} a \xrightarrow{q} a'' \to 0,$$

the morphism $\mathrm{Hom}(a,e) \xrightarrow{j^*} \mathrm{Hom}(a',e)$ is surjective if and only if the property (1) holds (here $j^*(f)) = f \circ j$). But the surjectivity of j^* is exactly the existence of the morphism g in diagram (2.2).

(2) If $\mathrm{Hom}(-,e)$ is exact, then applying it to the sequence (2.3), we get an exact sequence

$$0 \to \mathrm{Hom}(a'',e) \xrightarrow{p^*} \mathrm{Hom}(a,e) \xrightarrow{i^*} \mathrm{Hom}(e,e) \to 0.$$

Take a counterimage f of id_e in $\mathrm{Hom}(a,e)$, then $f \circ i = \mathrm{id}_e$, i.e., f splits the exact sequence (2.3).

Conversely, assuming that all sequences such as (2.3) split, we want to show that $\mathrm{Hom}(a,e) \xrightarrow{j^*} \mathrm{Hom}(a',e)$ is surjective.

Let $f \in \mathrm{Hom}(a',e)$ and let b be the pushout of (f,j),[c] so that there is a commutative diagram

$$
\begin{array}{ccccccc}
0 & \longrightarrow & a' & \xrightarrow{\ j\ } & a & & \\
 & & \downarrow f & & \downarrow g & & \\
0 & \longrightarrow & e & \xrightarrow{\ i\ } & b & \longrightarrow & b/e \longrightarrow 0.
\end{array}
$$

---

[b]We may say that g extends f.
[c]For the notion of pushout, see Appendix A.1.

Since j is a monomorphism, i is a monomorphism as well (see Lemma A.14), so we can complete i to an exact sequence by including the quotient b/e. By hypothesis this exact sequence splits, so that there is morphism $h : b \to e$ such that $h \circ g \circ j = f$. Thus $j^*(h \circ g) = f$. $\square$

Definition 2.5. An object e in an abelian category $\mathfrak{A}$ is said to be *injective* if $\mathrm{Hom}(-, e)$ is exact. The category $\mathfrak{A}$ is said to *have enough injectives* if every object in $\mathfrak{A}$ embeds into an injective object.

We are going to show that the category of abelian groups has enough injectives; indeed, *divisible groups* are injective abelian groups, and every abelian group can be injected into a divisible group. This will allow us to show that also the category of modules over a ring has enough injectives. We shall need the following result, which refers to the constructions made in Example 1.14. In particular, $H : \mathfrak{Ab} \to R\text{-}\mathbf{mod}$ will be the right adjoint to the forgetful functor $FF : R\text{-}\mathbf{mod} \to \mathfrak{Ab}$ given by

$$H(G) = \mathrm{Hom}_{\mathbb{Z}}(R, G).$$

Proposition 2.6. *Let R be a commutative ring with unity, and I an injective abelian group. Then $H(I) = \mathrm{Hom}_{\mathbb{Z}}(R, I)$ is an injective R-module.*

Proof. Let $M \to N$ be a monomorphism of R-modules. By the definition of adjunction, the following diagram commutes and its vertical arrows are isomorphisms:

$$
\begin{array}{ccc}
\mathrm{Hom}(N, H(I)) & \longrightarrow & \mathrm{Hom}(M, H(I)) \\
\simeq \downarrow & & \downarrow \simeq \\
\mathrm{Hom}(FF(N), I) & \longrightarrow\!\!\!\!\!\rightarrow & \mathrm{Hom}(FF(M), I)
\end{array}
$$

The arrow at the bottom is an epimorphism as I is injective (note that FF is an exact functor). Then the arrow at the top is an epimorphism as well, which means that $H(I)$ is an injective object. $\square$

Example 2.7. As a consequence of Corollary A.21 to Baer's Criterion in Appendix A.3, the injective objects in the category of abelian groups $\mathfrak{Ab}$ are the divisible groups, i.e., abelian groups G such that for every $g \in G$ and every nonzero $n \in \mathbb{Z}$ there is $h \in G$ such that $g = nh$ (apply the Corollary for $R = \mathbb{Z}$). For example, the additive group of rational numbers $\mathbb{Q}$ and the quotient group $\mathbb{Q}/\mathbb{Z}$ are divisible groups.

Now we prove that the category of abelian groups has enough injectives; in particular, we show that given an abelian group G, the group

$$I(G) = \prod_{\mathrm{Hom}_{\mathbb{Z}}(G,\mathbb{Q}/\mathbb{Z})} \mathbb{Q}/\mathbb{Z}$$

is injective, and that G injects into it. The fact that $I(G)$ is an injective group follows from the general fact that a product of injective objects is injective. Indeed, if $\{I_i\}$ is a family of injective objects, the product $\mathrm{Hom}_{\mathbb{Z}}(-, \prod_i I_i) = \prod_i \mathrm{Hom}_{\mathbb{Z}}(-, I_i)$ is an exact functor because an arbitrary product of epimorphisms is an epimorphism.

We check that if $G \neq 0$ the group $\mathrm{Hom}_{\mathbb{Z}}(G, \mathbb{Q}/\mathbb{Z})$ is nonzero, and, moreover, for any nonzero $g \in G$, there exists a morphism $\rho : G \to \mathbb{Q}/\mathbb{Z}$ such that $\rho(g) \neq 0$. The order of g is either infinite or an integer number $k \neq 1$. If the order is infinite, define the monomorphism of groups

$$i : \mathbb{Z} \to \langle g \rangle$$

$$1 \mapsto g,$$

where $\langle g \rangle$ is the cyclic group generated by g. Since $\mathbb{Q}/\mathbb{Z}$ is divisible, hence injective, for each $0 \neq [\frac{p}{q}] \in \mathbb{Q}/\mathbb{Z}$ there exists a morphism $\overline{\rho}_g^{[p/q]} : G \to \mathbb{Q}/\mathbb{Z}$ such that the diagram

$$
\begin{array}{ccc}
0 \longrightarrow \mathbb{Z} & \overset{i}{\longrightarrow} & G \\
\;\;\downarrow{\scriptstyle \rho_g^{[p/q]}} & \overset{\overline{\rho}_g^{[p/q]}}{\diagup} & \\
\mathbb{Q}/\mathbb{Z} & &
\end{array}
$$

commutes, where $\rho_g^{[p/q]}(1) = [p/q]$. Note that $\overline{\rho}_g^{[p/q]}(g) = [p/q] \neq 0$.

If $|g| = k$, the morphism

$$i : \mathbb{Z}_k \to \langle g \rangle$$

$$1 \mapsto g$$

is injective, and the nonzero morphism

$$\rho_g^{[h]} : \mathbb{Z}_k \to \mathbb{Q}/\mathbb{Z},$$

$$1 \mapsto [h],$$

where $h \in \mathbb{Q}$ is such that $k[h] = 0$ and $[h] \neq 0$, can be extended to a morphism $\overline{\rho}_g^{[h]} : G \to \mathbb{Q}/\mathbb{Z}$, according to the diagram

$$0 \longrightarrow \mathbb{Z}_k \overset{i}{\longrightarrow} G$$

$$\rho_g^{[h]} \downarrow \quad \overset{\overline{\rho}_g^{[h]}}{\swarrow}$$

$$\mathbb{Q}/\mathbb{Z}$$

Again, $\overline{\rho}_g^{[h]}(g) = \rho_g^{[h]}(1) = [h] \neq 0$.

Finally, we show that G injects into $I(G)$. Indeed, the homomorphism

$$e_G : G \to I(G)$$

$$g \mapsto \prod_\rho \rho(g),$$

where $\rho \in \mathrm{Hom}(G, \mathbb{Q}/\mathbb{Z})$, is injective, since, as we have just shown, for every $g \neq 0$ there exists a homomorphism ρ such that $\rho(g) \neq 0$.

Proposition 2.8. *Let R be a commutative ring R with unity. The category R-mod has enough injectives.*

Proof. The R-module $I = \mathrm{Hom}_{\mathbb{Z}}(R, \mathbb{Q}/\mathbb{Z})$ (see Example 1.14) is nonzero and injective by Proposition 2.6. If M is an R-module, we define

$$I(M) = \prod_{\mathrm{Hom}_R(M,I)} I.$$

Note that $\mathrm{Hom}_R(M, I) \simeq \mathrm{Hom}_{\mathbb{Z}}(M, \mathbb{Q}/\mathbb{Z})$ because of the adjunction between the functors FF and H of Example 1.14.

The homomorphism of R-modules

$$e_M : M \to I(M)$$

$$m \mapsto \prod_{\alpha \in \mathrm{Hom}_R(M,I)} \alpha(m)$$

is a monomorphism; indeed, if $m \in M$ is nonzero, there is $\alpha \in \mathrm{Hom}_R(M, I)$ such that $\alpha(m) \neq 0$. To show this one uses the isomorphism $\mathrm{Hom}_R(M, I) \simeq \mathrm{Hom}_{\mathbb{Z}}(M, \mathbb{Q}/\mathbb{Z})$; again in Example 2.7 we showed the existence of a morphism in the second group whose action on a given m is nonzero. $\square$

The notion of *resolution* of an object in an abelian category will be fundamental for the development of the theory of derived functors.

Definition 2.9. Given an object a in an abelian category $\mathfrak{A}$, a *resolution* of a is a pair $(L^\bullet, \epsilon)$, where $L^\bullet$ is a complex in $\mathfrak{A}$ indexed by the natural numbers, and ϵ is a morphism $a \to L^0$, such that the sequence

$$0 \to a \xrightarrow{\epsilon} L^0 \to L^1 \to L^2 \to \cdots$$

is exact.

If all objects I^i, $i \in \mathbb{N}$, are injective, we say that $(I^\bullet, \epsilon)$ is an *injective resolution* of a.

Proposition 2.10. *If $\mathfrak{A}$ has enough injectives, every object in $\mathfrak{A}$ has an injective resolution.*

Proof. An object a in $\mathfrak{A}$ embeds into an injective object e_0; let $q_0 = e_0/a$ be the quotient, so that there is an exact sequence $0 \to a \xrightarrow{i} e_0 \xrightarrow{p} q_0 \to 0$. Now we can embed q_0 into an injective object e_1, with quotient q_1. The kernel of the composition $e_0 \to e_1$ is a. Iterating this procedure, as shown in the diagram

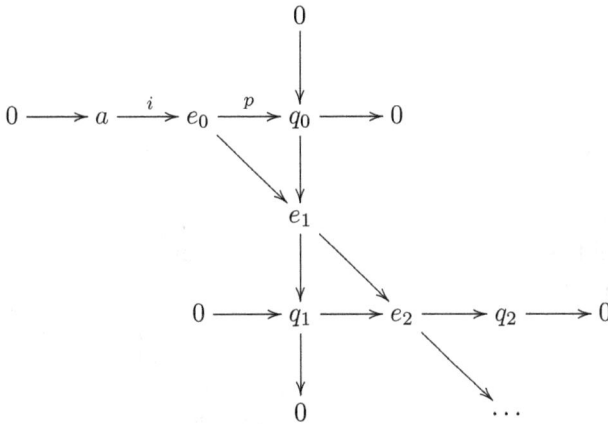

we build up an injective resolution. $\qquad\square$

Proposition 2.11 (Lifting property). *If $f : a \to a'$ is a morphism in an abelian category $\mathfrak{A}$, $L^\bullet$ is a resolution of a, and $(M^\bullet, d')$ is a complex of injective objects with a morphism $\eta : a' \to M^0$ such that $d'_0 \circ \eta = 0$, there exists a morphism of complexes $g : L^\bullet \to M^\bullet$, which lifts f, such that the diagram*

commutes. Any two such lifts are homotopic.

Proof. By induction, suppose that this has been done up to g_{n-1}, so that we have

$$L^{n-2} \xrightarrow{d_{n-2}} L^{n-1} \xrightarrow{d_{n-1}} L^n$$

$$\downarrow{g_{n-2}} \qquad \downarrow{g_{n-1}}$$

$$M^{n-2} \xrightarrow{d'_{n-2}} M^{n-1} \xrightarrow{d'_{n-1}} M^n$$

then we have

$$0 \longrightarrow L^{n-1}/\mathrm{im}\, d_{n-2} \xrightarrow{\ i\ } L^n$$

$$\downarrow{\bar{g}_{n-1}} \qquad \qquad \vdots\, g_n$$

$$M^{n-1}/\mathrm{im}\, d'_{n-2} \xrightarrow{\ j\ } M^n$$

Since M^n is injective there is g_n such that $g_n \cdot i = j \cdot \bar{g}_{n-1}$, and then, up to degree n, g is a morphism of complexes.

To start the induction we note that we have

$$0 \longrightarrow a \longrightarrow L^0$$

$$\downarrow{f} \qquad \vdots\, g_0$$

$$a' \xrightarrow{\ \eta\ } M^0$$

so g_0 exists because M^0 is injective.

Uniqueness up to homotopy. We prove the claim by induction. Suppose h is another lift, and that there is a homotopy k up to order n. We form the following diagram:

where

$$d'_{n-2} \circ k_{n-1} + k_n \circ d_{n-1} = g_{n-1} - h_{n-1}.$$

Given the morphism $t_n = g_n - h_n - d'_{n-1} \cdot k_n : L^n \to M^n$, we have

$$t_n \circ d_{n-1} = g_n \circ d_{n-1} - h_n \circ d_{n-1} - d'_{n-1} \circ k_n \circ d_{n-1}$$

$$= g_n \circ d_{n-1} - h_n \circ d_{n-1} - d'_{n-1}$$

$$\circ (g_{n-1} - h_{n-1} - d'_{n-2} \circ k_{n-1}) = 0.$$

So t_n factors through a morphism $t'_n : L^n / \operatorname{im} d_{n-1} \to M^n$, and since M^n is injective, and $L^n / \operatorname{im} d_{n-1} \to L^{n+1}$ is a monomorphism, there exists a morphism $k_{n+1} : L^{n+1} \to M^n$ such that

$$k_{n+1} \circ d_n + d'_{n-1} \circ k_n = t_n + d'_{n-1} \circ k_n = g_n - h_n,$$

i.e., we have a homotopy up to order $n + 1$.

To start the induction note that the morphism $t_0 = g_0 - h_0$ is zero on a, as

$$t_0 \circ \epsilon = (g_0 - h_0) \circ \epsilon = \eta \circ f - \eta \circ f = 0$$

so that we may form the diagram

$$
\begin{array}{ccc}
0 \longrightarrow L^0/a & \xrightarrow{\ d_0\ } & L^1 \\
\ \downarrow{\scriptstyle t'_0} & \swarrow {\scriptstyle k_1} & \\
M^0 & &
\end{array}
$$

Again, the morphism k_1 exists as M^0 is injective, and of course $k_1 \circ d_0 = g_0 - h_0$. $\qquad\square$

The following corollary is the key to a good definition of the derived functors.

Corollary 2.12. *Two injective resolutions $I^\bullet$, $J^\bullet$ of the same object in $\mathfrak{A}$ are homotopy equivalent.*

Proof. Since both $I^\bullet$ and $J^\bullet$ are injective resolutions, the identity $a \to a$ lifts to morphisms $h : I^\bullet \to J^\bullet$ and $g : J^\bullet \to I^\bullet$. Then both $h \circ g : I^\bullet \to I^\bullet$ and the identity morphism $I^\bullet \to I^\bullet$ lift the identity id_a, so that they are homotopic. For the same reason, $g \circ h$ is homotopic to $\operatorname{id}_{J^\bullet}$; cf. the diagram

$$
\begin{array}{ccc}
0 \longrightarrow a \longrightarrow & I^\bullet \\
\ \| & \ \uparrow{\scriptstyle g}\ \downarrow{\scriptstyle h} \\
0 \longrightarrow a \longrightarrow & J^\bullet
\end{array}
$$

$\qquad\square$

2.2. Right Derived Functors

As we already hinted, given a left exact functor, its right derived functors measure how far the functor is from being exact (cf. Theorem 2.17).

Let $F : \mathfrak{A} \to \mathfrak{B}$ be a left exact functor between abelian categories, and assume that $\mathfrak{A}$ has enough injectives. Fix an injective resolution $I^{\bullet}$ of every object $a \in \mathrm{Ob}(\mathfrak{A})$, and for every $i \geq 0$ define a functor

$$R^i F : \mathfrak{A} \to \mathfrak{B}$$

by letting

$$R^i F(a) = H^i(F(I^{\bullet})).$$

This is called the *i-th right derived functor of F*. Due to Corollary 2.12, these functors are independent of the choice of the injective resolutions up to isomorphisms of functors.

Actually, so far we defined these functors on the objects, and we need now to define them on morphisms. Given a morphism $f : a_1 \to a_2$ in $\mathfrak{A}$, consider injective resolutions I_1, I_2 of a_1, a_2, respectively. Due to Proposition 2.11, there is a lift $g : I_1^{\bullet} \to I_2^{\bullet}$ making the diagram

$$
\begin{array}{ccccc}
0 & \longrightarrow & a_1 & \longrightarrow & I_1^{\bullet} \\
& & \downarrow{\scriptstyle f} & & \downarrow{\scriptstyle g} \\
0 & \longrightarrow & a_2 & \longrightarrow & I_2^{\bullet}
\end{array}
$$

commutative. Applying the left exact functor F we obtain another commutative diagram

$$
\begin{array}{ccccc}
0 & \longrightarrow & F(a_1) & \longrightarrow & F(I_1^{\bullet}) \\
& & \downarrow{\scriptstyle F(f)} & & \downarrow{\scriptstyle F(g)} \\
0 & \longrightarrow & F(a_2) & \longrightarrow & F(I_2^{\bullet})
\end{array}
$$

This defines morphisms

$$H^i(F(I_1^{\bullet})) \to H^i(F(I_2^{\bullet})),$$

i.e., morphisms $R^i F(f) : R^i F(a_1) \to R^i F(a_2)$.

Remark 2.13.

- $R^0 F \simeq F$ as F is left exact.
- If a is injective, $R^i F(a) = 0$ for all $i > 0$, as $0 \to a \xrightarrow{\text{id}} a \to 0$ is an injective resolution.

Example 2.14. Given a commutative ring R and an R-module M, the covariant functor

$$\mathrm{Hom}_R(M, -) : R\text{-}\mathbf{mod} \to R\text{-}\mathbf{mod}$$

$$N \mapsto \mathrm{Hom}_R(M, N)$$

is additive and left exact, cf. Example 1.34. Its right derived functors are denoted $\mathrm{Ext}^i_R(M, -)$.

More generally, any object a in an abelian category $\mathfrak{A}$ with enough injectives defines a left exact functor

$$\mathrm{Hom}_{\mathfrak{A}}(a, -) : \mathfrak{A} \to \mathfrak{Ab};$$

its right derived functors are denoted $\mathrm{Ext}^i_{\mathfrak{A}}(a, -)$.

Example 2.15 (Lie algebra cohomology). A Lie algebra over a commutative ring R is an R-module $\mathfrak{g}$ with a skew bilinear operation

$$[-, -] : \mathfrak{g} \times \mathfrak{g} \to \mathfrak{g},$$

called *bracket*, which satisfies the *Jacobi identity*:

$$[x, [y, z]] + [z, [x, y]] + [y, [z, x]] = 0 \qquad \text{for all } x, y, z \in \mathfrak{g}.$$

A *representation* of $\mathfrak{g}$ is a pair (M, ρ), where M is an R-module, and ρ is an R-module morphism $\rho : \mathfrak{g} \to \mathrm{End}_R(M)$ such that

$$\rho([x, y]) = [\rho(x), \rho(y)],$$

where the bracket in the right-hand side is the commutator of endomorphisms. One defines the *invariant submodule* $M^{\mathfrak{g}}$ of M as

$$M^{\mathfrak{g}} = \{m \in M \mid \rho(x)(m) = 0 \text{ for all } x \in \mathfrak{g}\}.$$

Given representations (M, ρ), (M', ρ') of $\mathfrak{g}$, a morphism between them is an R-module morphism $f : M \to M'$ such that $f \circ \rho(x) = \rho'(x) \circ f$ for all $x \in \mathfrak{g}$. It is not difficult to show that the category $\mathrm{Rep}(\mathfrak{g})$ of representations of $\mathfrak{g}$ is abelian.

The assignment $(M, \rho) \mapsto M^{\mathfrak{g}}$ defines the *functor of invariants*

$$(-)^{\mathfrak{g}} : \mathrm{Rep}(\mathfrak{g}) \to R\text{-}\mathbf{mod}.$$

Quite clearly, this is a covariant additive functor. A few direct calculations also show that it is left exact. To derive this functor one needs to show that the category $\mathrm{Rep}(\mathfrak{g})$ has enough injectives. This is accomplished by means of the *universal enveloping algebra* $U(\mathfrak{g})$ of $\mathfrak{g}$ [33, 35, 64], which is an associative R-algebra defined as

$$U(\mathfrak{g}) = T(\mathfrak{g})/\mathfrak{a},$$

where $T(\mathfrak{g})$ is the tensor algebra of $\mathfrak{g}$ over R, and $\mathfrak{a}$ is the 2-sided ideal generated by elements of the form

$$x \otimes y - y \otimes x - [x, y].$$

Note that there is a natural monomorphism $i : \mathfrak{g} \to U(\mathfrak{g})$ satisfying

$$i([x, y]) = i(x)i(y) - i(y)i(x) \quad \text{for all } x, y \in \mathfrak{g}. \tag{2.4}$$

It turns out that $\mathrm{Rep}(\mathfrak{g})$ is isomorphic to the category of left $U(\mathfrak{g})$-modules [64, Corollary 7.3.4]. Let us check how this isomorphism is defined on objects. If M is a $U(\mathfrak{g})$-module, and $x \in \mathfrak{g}$ and $m \in M$, one defines a representation ρ by letting

$$\rho(x)(m) = i(x)m$$

for all $x \in \mathfrak{g}$ and $m \in M$. Conversely, if $u \in U(\mathfrak{g})$, we may assume that $u = [x_1 \otimes \cdots \otimes x_n]$ with $x_i \in \mathfrak{g}$; then let

$$um = \rho(x_1) \circ \cdots \circ \rho(x_n)(m).$$

This is well defined due to equation (2.4).

Now, every category of modules has enough injectives (cf. Proposition 2.8), so that the same is true for $\mathrm{Rep}(\mathfrak{g})$.[d]

The derived functors $R^i(-)^{\mathfrak{g}}$, applied to a representation (M, ρ), are the *Lie algebra cohomology groups* $H^i(\mathfrak{g}, M)$ of $\mathfrak{g}$ with coefficients in M (actually, they are R-modules).

2.3. Free Resolutions

In the case of the category of modules over a commutative ring, the Ext functors can also be computed using a resolution of the first, as opposed to the second, argument (this happens because Hom is actually a bifunctor, i.e., a functor depending on two arguments).

---

[d] Actually $U(\mathfrak{g})$ is not a commutative ring, however the proof of Proposition 2.8 works also in the noncommutative case.

Given an R-module M, a *free projective resolution* of M is a homology complex of free R-modules

$$\cdots \to F_n \xrightarrow{d_n} F_{n-1} \to \cdots \to F_1 \xrightarrow{d_1} F_0$$

together with a morphism $\epsilon : F_0 \to M$ such that the sequence

$$\cdots F_n \xrightarrow{d_n} F_{n-1} \to \cdots \to F_1 \xrightarrow{d_1} F_0 \xrightarrow{\epsilon} M \to 0$$

is exact. Such a resolution computes the derived functors of Hom with respect to the first argument, and these turn out to be the same Ext functors that one obtains by deriving with respect to the second argument. The computation of the Ext functors is usually easier in this way, because of the nicer properties of the free modules, and because free resolutions are often finite (i.e., they have a finite number of terms).

Proposition 2.16. *Let $F_\bullet$ be a free resolution of an R-module M. For every R-module N and every $i \geq 0$, there are natural isomorphisms*

$$H^i(\mathrm{Hom}_R(F_\bullet, N)) \simeq \mathrm{Ext}^i_R(M, N).$$

Proof. (Sketch) Let $0 \to N \to I^\bullet$ be an injective resolution of N, and consider the total complex $T^\bullet(F, I)$ of the double complex $\mathrm{Hom}_R(F_\bullet, I^\bullet)$, i.e.,

$$T^i(F, I) = \bigoplus_{i=m+n} \mathrm{Hom}_R(F_m, I^n)$$

with the obvious differential.[e] There are natural morphisms

$$\mathrm{Hom}_R(M, I^\bullet) \to T^\bullet(F, I) \leftarrow \mathrm{Hom}_R(F_\bullet, N)$$

---

[e]Given a double complex in an abelian category

$$K = \bigoplus_{p,q \in \mathbb{Z}} K^{p,q}$$

with differentials

$$d_1 : K^{p,q} \to K^{p+1,q}, \qquad d_2 : K^{p,q} \to K^{p,q+1},$$

one defines the total complex as

$$T^n(K) = \bigoplus_{p+q=n} K^{p,q}.$$

Usually two situations arise, i.e., the two differentials either anticommute or commute. In the first case, one defines the differential of the total complex as $d = d_1 + d_2$, so that $d^2 = 0$. In the second case, to achieve $d^2 = 0$ one sets $d = d_1 + (-1)^p d_2$ when acting on $K^{p,q}$.

which turn out to be *quasi-isomorphisms*, i.e., the morphisms they induce between the cohomology modules of the respective complexes are isomorphisms. For a proof of this fact, see, e.g., [64, Section II.7].

As a result, we have

$$H^i(\operatorname{Hom}_R(F_\bullet, N)) \simeq H^i(\operatorname{Hom}_R(M, I^\bullet)) = \operatorname{Ext}^i_R(M, N). \qquad \square$$

Examples of the computation of the Ext groups by means of free resolutions are given in the exercises at the end of this chapter.

Proposition 2.16 foreshadows the fact that in some abelian categories the Ext functors can be computed as derived functors of the Hom functor with respect to the first argument. This is actually the content of Theorem 2.7.6 in [64], which we have referred for a proof of Proposition 2.16; we shall prove this result in Section 5.6.1 using spectral sequences.

2.4. Long Exact Sequence of a Derived Functor

Again, the right derived functors of a left exact functor F measure how far F is from being exact. The following theorem makes precise sense of this fact. It is essentially based on Proposition 1.23, together with some technical results in homological algebra.

Theorem 2.17. *Let*

$$0 \to a' \xrightarrow{i} a \xrightarrow{p} a'' \to 0$$

be an exact sequence in an abelian category $\mathfrak{A}$ with enough injectives, and $F : \mathfrak{A} \to \mathfrak{B}$ a left exact functor, where the category $\mathfrak{B}$ is abelian as well. There are morphisms $\delta_n : R^n F(a'') \to R^{n+1} F(a')$ for $n \geq 0$ such that the sequence

$$0 \to F(a') \xrightarrow{F(i)} F(a) \xrightarrow{F(p)} F(a'') \xrightarrow{\delta_0} R^1 F(a')$$

$$\xrightarrow{R^1 F(i)} R^1 F(a) \xrightarrow{R^1 F(p)} R^1 F(a'') \xrightarrow{\delta_1} \cdots$$

is exact.

The proof of this result will need two preliminary lemmas.

Lemma 2.18 (Horseshoe Lemma). *If*

$$0 \to a' \to a \to a'' \to 0$$

is an exact sequence in an abelian category $\mathfrak{A}$ with enough injectives, there exist injective resolutions $I^\bullet, J^\bullet, K^\bullet$ of a', a, a'', respectively, that fit into a *commutative diagram with exact rows*

$$
\begin{array}{ccccccccc}
& & 0 & & 0 & & 0 & & \\
& & \downarrow & & \downarrow & & \downarrow & & \\
0 & \longrightarrow & a' & \longrightarrow & a & \longrightarrow & a'' & \longrightarrow & 0 \\
& & \downarrow & & \downarrow & & \downarrow & & \\
0 & \longrightarrow & I^\bullet & \longrightarrow & J^\bullet & \longrightarrow & K^\bullet & \longrightarrow & 0
\end{array}
\qquad (2.5)
$$

Proof. The proof uses induction on the degree of the terms in the three resolutions. Choose injective resolutions $I^\bullet$ and $K^\bullet$ of a and a'', set $J^\bullet = I^\bullet \oplus K^\bullet$, and form the diagram

$$
\begin{array}{ccccccccc}
& & 0 & & 0 & & 0 & & \\
& & \downarrow & & \downarrow & & \downarrow & & \\
0 & \longrightarrow & a' & \longrightarrow & a & \longrightarrow & a'' & \longrightarrow & 0 \\
& & \downarrow & & \downarrow & & \downarrow & & \\
0 & \longrightarrow & I^0 & \longrightarrow & J^0 & \longrightarrow & K^0 & \longrightarrow & 0
\end{array}
$$

The dotted arrow exists because I^0 is injective. A morphism $a \to K^0$ is defined by the composition $a \to a'' \to K^0$; together with the morphism $a \to I^0$ this gives a morphism $a \to J^0$ which evidently makes the diagram commutative. The last row is evidently exact. This proves the induction basis.

Now for the induction step.[f] Let us form the diagram[g]

$$
\begin{array}{ccccccccc}
0 & \longrightarrow & I^{n-1} & \longrightarrow & J^{n-1} & \longrightarrow & K^{n-1} & \longrightarrow & 0 \qquad (2.6) \\
& & \downarrow{\scriptstyle d'_{n-1}} & & \downarrow{\scriptstyle d_{n-1}} & & \downarrow{\scriptstyle d''_{n-1}} & & \\
0 & \longrightarrow & I^n & \longrightarrow & J^n & \longrightarrow & K^n & \longrightarrow & 0 \\
& & \downarrow & & \downarrow & & \downarrow & & \\
0 & \dashrightarrow & \operatorname{coker} d'_{n-1} & \longrightarrow & \operatorname{coker} d_{n-1} & \longrightarrow & \operatorname{coker} d''_{n-1} & \longrightarrow & 0 \\
& & \downarrow & & \downarrow & & \downarrow & & \\
0 & \longrightarrow & I^{n+1} & \longrightarrow & J^{n+1} & \longrightarrow & K^{n+1} & \longrightarrow & 0
\end{array}
$$

---

[f]One might object that since $J^\bullet$ is the direct sum of $I^\bullet$ and $K^\bullet$, morphisms $J^n \to J^{n+1}$ always exist, however in general they would not make the diagram (2.5) commutative.
[g]Let us note that, if a complex $(C^\bullet, d)$ is exact, for any n we may split it in two exact sequences

$$
\cdots \to C^{n-1} \xrightarrow{d_{n-1}} C^n \to \operatorname{coker} d_{n-1} \to 0, \qquad 0 \to \operatorname{coker} d_{n-1} \to C^{n+1} \to \cdots
$$

This is what we have done to form the columns of the diagram (2.6).

where the sequence in the third line exists by the Snake Lemma (see Appendix A.2) — but we have to prove that the morphism coker $d'_{n-1} \to$ coker d_{n-1} is injective. However, one can check that the retraction $J^n \to I^n$, which exists as the second line is split exact, defines a retraction of coker $d'_{n-1} \to$ coker d_{n-1} (one needs to use the fact that the first line is also split exact). As a consequence, coker $d'_{n-1} \to$ coker d_{n-1} is injective.

Finally, the morphism coker $d_{n-1} \to J^{n+1}$ is defined as before.   □

The name of the lemma comes from the fact that at first one chooses the resolutions $I^{\bullet}$ and $K^{\bullet}$; the diagram containing them has the shape of a horseshoe.

Lemma 2.19. *If* $F : \mathfrak{A} \to \mathfrak{B}$ *is a left exact functor between abelian categories and*

$$0 \to a' \to a \xrightarrow{p} a'' \to 0$$

is a split exact sequence in $\mathfrak{A}$, *then*

$$0 \to F(a') \to F(a) \xrightarrow{F(p)} F(a'') \to 0$$

splits in $\mathfrak{B}$ (*and is exact*).

Proof. The sequence $0 \to F(a') \to F(a) \xrightarrow{F(p)} F(a'')$ is exact as F is left exact. Now, if s is a section of p, the morphism $F(s) : F(a'') \to F(a)$ is a section of $F(p)$. Note that this implies that $F(p)$ is surjective.   □

Proof of Theorem 2.17. The second horizontal exact sequence in the commutative diagram (2.5) splits. By Lemma 2.19

$$0 \to F(I^{\bullet}) \to F(J^{\bullet}) \to F(K^{\bullet}) \to 0$$

is a split, hence exact sequence. The existence of the long exact sequence of derived functors follows from Proposition 1.23.   □

Ext groups and extensions. The Ext^1 groups in an abelian category $\mathfrak{A}$ with enough injectives classify the *extensions* in $\mathfrak{A}$, in a sense that we briefly describe now. Let

$$(A) \qquad 0 \to a \to b \to c \to 0 \qquad (2.7)$$

be an exact sequence in $\mathfrak{A}$. One says that b is an *extension* of c by a. If we apply the functor $\mathrm{Hom}_{\mathfrak{A}}(c, -)$ to the sequence (2.7), the resulting long exact sequence contains the connecting morphism

$$\mathrm{Hom}_{\mathfrak{A}}(c,c) \xrightarrow{\delta} \mathrm{Ext}^1_{\mathfrak{A}}(c,a)$$

and we may associate the class $e(A) = \delta(\mathrm{id}_c) \in \mathrm{Ext}^1_{\mathfrak{A}}(c,a)$ to the extension (2.7).

Another extension of c by a, say

$$(B) \qquad 0 \to a \to b' \to c \to 0,$$

is *equivalent* to (A) if there exists a morphism $b \to b'$ such that the diagram

commutes (if such a morphism exists, it is necessarily an isomorphism by the Five Lemma[h]).

One says that an extension as in (2.7) is *split* if it is equivalent to the trivial extension,

$$0 \to a \to a \oplus c \to c \to 0,$$

i.e., if the exact sequence (2.7) splits (see Definition 2.1).

Remark 2.20. The naturality of the Ext functors implies that two equivalent extensions yield the same element in the Ext group, i.e., if (A) and (B) are equivalent, then $e(A) = e(B)$.

Exercise 2.21. Check that an extension as in (2.7) is split if and only if it corresponds to the zero element in $\mathrm{Ext}^1_{\mathfrak{A}}(c,a)$.

For given objects c and a in $\mathfrak{A}$, let $E(c,a)$ be the set of equivalence classes of extensions of c by a.

---

[h]The strong form of the Five Lemma states that if a commutative diagram in an abelian category

is given, where the rows are exact, and the morphisms f_1, f_2, g_1, and g_2 are isomorphisms, then h is an isomorphism as well. This may be proved using the Snake Lemma, see Appendix A.2.

Proposition 2.22. *There is a set-theoretic bijection* $\mathrm{Ext}^1_{\mathfrak{A}}(c, a) \to E(c, a)$. *Under this correspondence, the split extension corresponds to the zero element in* $\mathrm{Ext}^1_{\mathfrak{A}}(c, a)$.

Proof. (Sketch) Fix $\xi \in \mathrm{Ext}^1_{\mathfrak{A}}(c, a)$. Embed a into an injective object I of $\mathfrak{A}$, and write the exact sequence

$$0 \to a \xrightarrow{i} I \to q \to 0.$$

Since $\mathrm{Ext}^1_{\mathfrak{A}}(c, I) = 0$ as I is injective, by applying $\mathrm{Hom}_{\mathfrak{A}}(c, -)$ to this exact sequence we obtain

$$\mathrm{Hom}_{\mathfrak{A}}(c, I) \to \mathrm{Hom}_{\mathfrak{A}}(c, q) \xrightarrow{\delta} \mathrm{Ext}^1_{\mathfrak{A}}(c, a) \to 0.$$

Pick $f \in \mathrm{Hom}_{\mathfrak{A}}(c, q)$ such that $\delta(f) = \xi$, and take the fibred product (see Appendix A.1) $p = c \times_q I$. At this stage we have a diagram

$$
\begin{array}{ccccc}
p & \longrightarrow & c & \longrightarrow & 0 \\
\downarrow & & \downarrow {\scriptstyle f} & & \\
0 \longrightarrow a & \xrightarrow{\ i\ } & I & \longrightarrow q & \longrightarrow 0
\end{array}
$$

Now the composition $a \xrightarrow{(0,i)} c \oplus I \to q$ is zero so that there is a morphism $a \to p$ (as $p = \ker(c \oplus I \to q)$), and the previous diagram can be completed to

$$
\begin{array}{ccccccc}
0 \dashrightarrow a & \longrightarrow & p & \longrightarrow & c & \longrightarrow & \to 0 \\
\| & & \downarrow & & \downarrow {\scriptstyle f} & & \\
0 \longrightarrow a & \xrightarrow{\ i\ } & I & \longrightarrow & q & \longrightarrow & 0
\end{array}
$$

We leave the reader to check that

(1) the first row in the last diagram is exact,[i] and
(2) a different choice of the morphism f produces an equivalent extension.

This establishes the map in the claim. Remark 2.20 asserts that there exists a map in the opposite direction. It is not difficult to check that the two maps are one the inverse of the other. For a detailed proof, the reader may consult [30].

    The second claim is the content of Exercise 2.21.   □

---

[i] One uses Lemmas A.13 and A.14.

2.5. Acyclic Resolutions

Right derived functors are defined by means of injective resolutions; however this is often impractical. It turns out that the right derived functors can be computed by using a wider class of resolutions. This is often very useful in practice.

Given abelian categories $\mathfrak{A}$ and $\mathfrak{B}$, with $\mathfrak{A}$ having enough injectives, and a left exact functor $F : \mathfrak{A} \to \mathfrak{B}$, we say that an object a in $\mathfrak{A}$ is F-*acyclic* if $R^i F(a) = 0$ for $i > 0$. Injective objects in $\mathfrak{A}$ are tautologically acyclic for any left exact functor on $\mathfrak{A}$. Moreover, we say that a resolution $L^\bullet$ of a is an F-*acyclic resolution* if all L^i are F-acyclic.

Theorem 2.23. *If* $0 \to a \to L^\bullet$ *is an* F-*acyclic resolution, there is a natural isomorphism* $H^i(F(L^\bullet)) \to R^i F(a)$ *for all* $i \geq 0$. *Natural means that if* $a \xrightarrow{f} a'$ *is a morphism in* $\mathfrak{A}$, *and* $L^\bullet$ *and* $M^\bullet$ *are* F-*acyclic resolutions of* a *and* a', *respectively, with a morphism* $g^\bullet : L^\bullet \to M^\bullet$ *that lifts* f, *then the induced morphisms* $H^i(F(L^\bullet)) \to H^i(F(M^\bullet))$ *make the diagrams*

$$
\begin{array}{ccc}
H^i(F(L^\bullet)) & \longrightarrow & R^i F(a) \\
{\scriptstyle H^i(F(g))}\downarrow & & \downarrow{\scriptstyle R^i F(f)} \\
H^i(F(M^\bullet)) & \longrightarrow & R^i F(a')
\end{array}
\tag{2.8}
$$

commutative.

Proof. By Proposition 2.11, if $I^\bullet$ is an injective resolution of a, there is a morphism of resolutions $L^\bullet \to I^\bullet$ inducing $H^i(F(L^\bullet)) \to R^i F(a)$, whether $L^\bullet$ is acyclic or not. We prove that when $L^\bullet$ is acyclic, this is an isomorphism.

The resolution $0 \to a \to L^\bullet$ splits into

$$
0 \to a \to L^0 \to J^1 \to 0
$$

$$
0 \to J^1 \to L^1 \to J^2 \to 0
$$

$$
\vdots
$$

$$
0 \to J^{i-1} \to L^{i-1} \to J^i \to 0
$$

$$
\vdots
$$

(2.9)

As the objects L^i are F-acyclic, by applying F to the first sequence we obtain

$$0 \to F(a) \to F(L^0) \to F(J^1) \to R^1F(a) \to 0$$
$$0 \to R^1F(J^1) \to R^2F(a) \to 0$$
$$\vdots$$
$$0 \to R^{i-1}F(J^1) \to R^iF(a) \to 0$$
$$\vdots$$

In the same way, by appyling F to the sequence (2.9) we obtain

$$0 \to F(J^{i-1}) \to F(L^{i-1}) \to F(J^i) \to R^1F(J^{i-1}) \to 0$$
$$0 \to R^1F(J^i) \to R^2F(J^{i-1}) \to 0$$
$$\vdots \qquad\qquad\qquad\qquad (2.10)$$
$$0 \to R^{j-1}F(J^i) \to R^jF(J^{i-1}) \to 0$$
$$\vdots$$

This produces isomorphisms

$$R^iF(a) \simeq R^{i-1}F(J^1) \simeq \cdots \simeq R^1F(J^{i-1}).$$

Moreover, from (2.10) we have

$$R^1F(J^{i-1}) \simeq \frac{F(J^i)}{\operatorname{im} F(L^{i-1})} \simeq \frac{\ker(F(L^i) \to F(J^{i+1}))}{\operatorname{im} F(L^{i-1})}.$$

Since the morphism $F(J^{i+1}) \to F(L^{i+1})$ is injective, we also have

$$R^1F(J^{i-1}) \simeq \frac{\ker(F(L^i) \to F(L^{i+1}))}{\operatorname{im} F(L^{i-1})} = H^i(F(L^\bullet)).$$

Putting all this together we get $R^iF(a) \simeq H^i(F(L^\bullet))$, thus proving the first claim.

For the second claim, by repeated use of the Lifting Property (Proposition 2.11) and by the fact that $\mathfrak{A}$ has enough injectives, one can introduce injective resolutions $I^\bullet$ of a and $J^\bullet$ of a' which fit into a commutative diagram

$$\begin{array}{ccc} L^\bullet & \longrightarrow & I^\bullet \\ \downarrow & & \downarrow \\ M^\bullet & \longrightarrow & J^\bullet \end{array}$$

Applying the functor F and taking cohomology we obtain the diagram (2.8). $\qquad\square$

This result is sometimes called the *Abstract de Rham Theorem*. Indeed, this generalizes the usual de Rham Theorem (our Theorem 4.41).

2.6. δ-functors

Derived functors enjoy some universal properties, which are very useful to prove uniqueness results. We formalize this fact by introducing the more general notion of δ-*functor* [22, 28, 59, 64].

Definition 2.24. A δ-functor between two abelian categories $\mathfrak{A}$ and $\mathfrak{B}$ is a collection of functors $\{T^i : \mathfrak{A} \to \mathfrak{B}, i \geq 0\}$, and for every exact sequence

$$0 \to a' \to a \to a'' \to 0$$

in $\mathfrak{A}$, morphisms $\delta_i : T^i(a'') \to T^{i+1}(a')$ for each $i \geq 0$, such that:

- there is a long exact sequence

$$0 \to T^0(a') \to T^0(a) \to T^0(a'') \xrightarrow{\delta_0} T^1(a') \cdots$$

$$\cdots \to T^i(a') \to T^i(a) \to T^i(a'') \xrightarrow{\delta_i} T^{i+1}(a') \to \cdots$$

- for each morphism of short exact sequences

$$
\begin{array}{ccccccccc}
0 & \longrightarrow & a' & \longrightarrow & a & \longrightarrow & a'' & \longrightarrow & 0 \\
& & \downarrow & & \downarrow & & \downarrow & & \\
0 & \longrightarrow & b' & \longrightarrow & b & \longrightarrow & b'' & \longrightarrow & 0
\end{array}
$$

there are commutative diagrams

$$
\begin{array}{ccc}
T^i(a'') & \xrightarrow{\delta_i} & T^{i+1}(a') \\
\downarrow & & \downarrow \\
T^i(b'') & \xrightarrow{\delta_i'} & T^{i+1}(b')
\end{array}
$$

We shall write a δ-functor as a collection of pairs $\{T^i, \delta_i\}$, but it is understood that the morphisms δ_i depend on the choice of the exact sequence. Note that if $\{T^i, \delta_i\}$ is a δ-functor, then T^0 is left-exact.

Definition 2.25. A functor $F : \mathfrak{A} \to \mathfrak{B}$ of abelian categories is *effaceable* if for every object a in $\mathfrak{A}$ there is a monomorphism $g : a \to b$ in $\mathfrak{A}$ such that $F(g) = 0$.

Note that, if $\mathfrak{A}$ has enough injectives, b can be embedded into an injective object, and therefore b itself can be assumed to be injective.

The following lemma establishes a useful criterion to check whether a functor is effaceable.

Lemma 2.26. *If $\mathfrak{A}$ has enough injectives, and $F(I) = 0$ for every injective object I in $\mathfrak{A}$, then F is effaceable.*

Proof. If $g : a \to I$ is the embedding of a into an injective object, then $F(g) = 0$ as $F(I) = 0$. $\qquad\square$

Definition 2.27. A δ-functor $\{T^i, \delta_i\}$ from $\mathfrak{A}$ to $\mathfrak{B}$ is *universal* if for any other δ-functor $\{S^i, \sigma_i\}$ from $\mathfrak{A}$ to $\mathfrak{B}$, with a morphism $f^0 : T^0 \to S^0$, there are morphisms $f^i : T^i \to S^i$, $i \geq 1$, such that for all exact sequences $0 \to a' \to a \to a'' \to 0$ the diagram

$$
\begin{array}{ccc}
T^i(a'') & \xrightarrow{\;\delta_i\;} & T^{i+1}(a') \\
{\scriptstyle f^i}\big\downarrow & & \big\downarrow{\scriptstyle f^{i+1}} \\
S^i(a'') & \xrightarrow{\;\sigma_i\;} & S^{i+1}(a')
\end{array}
$$

commutes.

Remark 2.28. Let $F : \mathfrak{A} \to \mathfrak{B}$ be a covariant left exact functor; any universal δ-functor $\{T^i, \delta_i\}$ such that $T^0 \simeq F$ is unique up to unique isomorphism.

Theorem 2.29. *A δ-functor $\{T^i, \delta_i\}$ from $\mathfrak{A}$ to $\mathfrak{B}$, such that for $i > 0$ the functor T^i is effaceable, is universal.*

Proof. Let $\{S^i, \sigma_i\}$ be another δ-functor, with a morphism $f^0 : T^0 \to S^0$. We define $f^1 : T^1 \to S^1$. Let $0 \to a' \xrightarrow{g} a \to a'' \to 0$ be an exact sequence in $\mathfrak{A}$ such that $T^1(g) = 0$, and form the diagram

$$
\begin{array}{ccccccc}
T^0(a) & \longrightarrow & T^0(a'') & \xrightarrow{\;\delta_0\;} & T^1(a') & \xrightarrow{\;0\;} & T^1(a) \\
{\scriptstyle f_a^0}\big\downarrow & & {\scriptstyle f_a^0}\big\downarrow & & \big\downarrow{\scriptstyle f_a^1} & & \\
S^0(a) & \longrightarrow & S^0(a'') & \xrightarrow{\;\sigma_0\;} & S^1(a') & &
\end{array}
$$

Since $T^1(g) = 0$, δ_0 is surjective and this defines the morphism f_a^1. Now one needs to show that f^1 is actually a morphism of functors, i.e., it does not depend on the choices made. This is easy to do when we assume that $\mathfrak{A}$

has enough injectives, so that, as we noted earlier, we may assume a to be injective. If b is another choice for a, in principle we get another morphism f_b^1, but since b is injective there is a morphism $a \to b$. We may form the commutative diagram

Some diagram chasing, which in particular uses the fact that the morphism $\delta_{0,a}$ is surjective, shows that $f_a^1 = f_b^1$. Therefore f^1 is a well-defined morphism of functors. This is true also for general $\mathfrak{A}$, although this is a little more difficult to prove.

The higher order morphisms f^i, $i \geq 2$, are recursively defined in the same way. $\qquad\square$

Corollary 2.30. *Let $F : \mathfrak{A} \to \mathfrak{B}$ be a left exact functor, where $\mathfrak{A}$ and $\mathfrak{B}$ are abelian categories, and $\mathfrak{A}$ has enough injectives. Then the functors $R^i F$, together with the morphisms defined in Theorem 2.17, make up a universal δ-functor. As a consequence, if $\{T^i\}$ is a universal δ-functor from $\mathfrak{A}$ to $\mathfrak{B}$, such that $T^0 \simeq F$, then $T^i \simeq R^i F$ for all $i \geq 0$.*

Proof. By Remark 2.13 and Lemma 2.26 the functors $R^i F$, $i > 0$, are effaceable. Then the first claim follows from Theorem 2.29. The second claim follows from Remark 2.28. $\qquad\square$

Example 2.31 (Chevalley–Eilenberg cohomology of a Lie algebra).
Let $\mathfrak{g}$ be a Lie algebra over a commutative ring R (cf. Example 2.15). We give another description of the cohomology theory associated with it, and use Corollary 2.30 to show that it coincides with the one described in Example 2.15. We assume that $\mathfrak{g}$ is free over R, and, after fixing a representation (M, ρ) of $\mathfrak{g}$, define the Chevalley–Eilenberg complex of $\mathfrak{g}$

with coefficients in M as [13]

$$C^p(\mathfrak{g}, M) = \operatorname{Hom}_R(\Lambda^p \mathfrak{g}, M), \quad p \geq 0,$$

where $\Lambda^p \mathfrak{g}$ is the pth exterior power of $\mathfrak{g}$.[j] A differential

$$d : C^p(\mathfrak{g}, M) \to C^{p+1}(\mathfrak{g}, M)$$

is defined by letting

$$(d\xi)(x_1, \ldots, x_{p+1}) = \sum_{i=1}^{p+1} (-1)^{i-1} \rho(x_i)(\xi(x_1, \ldots, \hat{x}_i, \ldots, x_{p+1}))$$

$$+ \sum_{i<j} (-1)^{i+j} \xi([x_i, x_j], \ldots, \hat{x}_i, \ldots, \hat{x}_j, \ldots, x_{p+1}),$$

where $\xi \in C^p(\mathfrak{g}, M)$, $x_1, \ldots, x_{p+1} \in \mathfrak{g}$, and a caret denotes omission. Note that in particular

$$d : (C^0(\mathfrak{g}, M) = M) \to (C^1(\mathfrak{g}, M) = \operatorname{Hom}_R(\mathfrak{g}, M))$$

is defined as

$$(dm)(x) = \rho(x)(m) \tag{2.11}$$

for $x \in \mathfrak{g}$ and $m \in M$.

We denote by $H_{CE}^i(\mathfrak{g}, M)$ the cohomology groups of the complex $C^\bullet(\mathfrak{g}, M)$. By (2.11), we have

$$H_{CE}^0(\mathfrak{g}, M) = M^{\mathfrak{g}}; \tag{2.12}$$

that is, $H_{CE}^0(\mathfrak{g}, -)$, as a functor $\operatorname{Rep}(\mathfrak{g}) \to R\text{-}\mathbf{mod}$, coincides with the functor of invariants $(-)^{\mathfrak{g}}$ defined in Example 2.15. More generally, the groups $H_{CE}^i(\mathfrak{g}, M)$ coincide with the groups $H^i(\mathfrak{g}, M)$ there defined. To prove this we shall use Corollary 2.30. In view of the isomorphism (2.12), what we need to show is that the collection of functors

$$H_{CE}^i(\mathfrak{g}, -) : \operatorname{Rep}(\mathfrak{g}) \to R\text{-}\mathbf{mod}$$

makes up a universal δ-functor. The existence of the connecting morphisms and of the long exact cohomology sequence, and their naturality, follow from the general cohomology theory so far developed. In view of Lemma 2.26,

---

[j]When $\mathfrak{g}$ is also finitely generated over R, we may write the Chevalley–Eilenberg complex as

$$C^p(\mathfrak{g}, M) = M \otimes_R \Lambda^p \mathfrak{g}^*, \quad p \geq 0.$$

to prove that the δ-functor so defined is universal it is enough to show that $H^i_{CE}(\mathfrak{g}, I) = 0$ for $i > 0$ whenever I is an injective object in $\mathrm{Rep}(\mathfrak{g})$. The remaining part of this Example will be devoted to proving that fact.

Consider the universal enveloping algebra $U(\mathfrak{g})$. If $\mathfrak{a}$ is the ideal generated by the image of the natural monomorphism $i : \mathfrak{g} \to U(\mathfrak{g})$, one has $U(\mathfrak{g})/\mathfrak{a} \simeq R$. The projection $\epsilon : U(\mathfrak{g}) \to R$ may be extended to a resolution of R by free $U(\mathfrak{g})$-modules.[k] Indeed one defines morphisms

$$\partial_p : U(\mathfrak{g}) \otimes_R \Lambda^p \mathfrak{g} \to U(\mathfrak{g}) \otimes_R \Lambda^{p-1} \mathfrak{g}$$

$$\partial_p(u \otimes x_1 \wedge \cdots \wedge x_p) = \sum_{j=1}^{p} (-1)^{j-1} u x_j \otimes x_1 \wedge \cdots \wedge \hat{x}_j \wedge \cdots \wedge x_p$$

$$+ \sum_{1 \le j < k \le p} (-1)^{j+k} u \otimes [x_j, x_k] \wedge x_1 \wedge \cdots \wedge \hat{x}_j$$

$$\wedge \cdots \wedge \hat{x}_k \wedge \cdots \wedge x_p$$

for $u \in U(\mathfrak{g})$ and $x_1, \ldots, x_p \in \mathfrak{g}$; in particular, $\partial_1(u \otimes x) = ux$. It turns out that

$$\cdots \to U(\mathfrak{g}) \otimes_R \Lambda^2 \mathfrak{g} \xrightarrow{\partial_2} U(\mathfrak{g}) \otimes_R \mathfrak{g} \xrightarrow{\partial_1} U(\mathfrak{g}) \xrightarrow{\epsilon} R \to 0$$

is an exact complex, that is, $U(\mathfrak{g}) \otimes_R \Lambda^\bullet \mathfrak{g}$ is a resolution of R (we refer the reader to [64, Theorem 7.7.2] for a proof). As I is an injective object in $\mathrm{Rep}(\mathfrak{g}) \simeq U(\mathfrak{g})\text{-}\mathbf{mod}$, the functor $\mathrm{Hom}_{U(\mathfrak{g})}(-, I)$ is exact, so that the complex $\mathrm{Hom}_{U(\mathfrak{g})}(U(\mathfrak{g}) \otimes_R \Lambda^\bullet \mathfrak{g}, I)$ is exact in positive degree. Now one has

$$\mathrm{Hom}_{U(\mathfrak{g})}(U(\mathfrak{g}) \otimes_R \Lambda^\bullet \mathfrak{g}, I) \simeq \mathrm{Hom}_R(\Lambda^\bullet \mathfrak{g}, I) = C^\bullet(\mathfrak{g}, I);$$

thus we have proved that the complex $C^\bullet(\mathfrak{g}, I)$ is acyclic, i.e., $H^p_{CE}(\mathfrak{g}, I) = 0$ for $p > 0$.

2.7. Left Derived Functors

So far we have developed the theory of right derived functors of left exact functors defined on an abelian category that has enough injectives. A dual theory can be developed to build *left derived functors* of *right exact functors*, using *projective resolutions* instead of injective ones. We shall give in this section a cursory introduction to this theory.

---

[k]When $\mathfrak{g}$ is free over R, its universal enveloping algebra is a free module over R; this is a consequence of the Poincaré–Birkhoff–Witt Theorem, see, e.g., [35, Section 5.2], or [64].

In analogy with the notion of injective object, an object P in an abelian category $\mathfrak{A}$ is said to be *projective* if one of the following three equivalent conditions is verified:

(1) the functor $\mathrm{Hom}_{\mathfrak{A}}(P, -) : \mathfrak{A} \to \mathfrak{Ab}$ is exact;
(2) if $f : a \to b$ is an epimorphism in $\mathfrak{A}$, any morphism $P \to b$ factors through f:

$$
\begin{array}{ccc}
 & & P \\
 & \nearrow & \downarrow \\
a & \xrightarrow{\ f\ } b & \longrightarrow 0
\end{array}
$$

(3) every exact sequence in $\mathfrak{A}$

$$0 \to a \to b \to P \to 0$$

splits.

A *projective resolution* of an object a in $\mathfrak{A}$ is a homology complex of projective objects

$$\cdots \to P_n \xrightarrow{d_n} P_{n-1} \to \cdots \to P_1 \xrightarrow{d_1} P_0$$

together with a morphism $\epsilon : P_0 \to a$ such that the sequence

$$\cdots P_n \xrightarrow{d_n} P_{n-1} \to \cdots \to P_1 \xrightarrow{d_1} P_0 \xrightarrow{\epsilon} a \to 0$$

is exact. One says that $\mathfrak{A}$ has *enough projectives* if every object is a quotient of a projective object. In analogy with Proposition 2.10, one proves that in a category with enough projectives, every object has a projective resolution.

Example 2.32. Free R-modules are projective; more generally, an R-module is projective if and only if it is a direct summand of a free module [30, 64].

If $F : \mathfrak{A} \to \mathfrak{B}$, where $\mathfrak{A}$ and $\mathfrak{B}$ are abelian categories and $\mathfrak{A}$ has enough projectives, is a right exact functor, one defines the *left derived functors* of F as

$$L_i F(a) = H_i(F(P_\bullet)),$$

where $H_i(F(P_\bullet))$ are the *homology* objects of the complex $F(P_\bullet)$, i.e.,

$$H_i(F(P_\bullet)) = \frac{\ker F(P_i) \to F(P_{i-1})}{\mathrm{im}\, F(P_{i+1}) \to F(P_i)}.$$

There is an isomorphism $L_0 F \simeq F$. If P is a projective object, then $L_i F(P) = 0$ for all $i > 0$ and every right exact functor F.

Example 2.33. The category of R-modules has enough projectives [30, 64]. Given an R-module M, the tensor product functor

$$M \otimes_R - : R\text{-}\mathbf{mod} \to R\text{-}\mathbf{mod}$$

$$N \mapsto M \otimes_R N$$

is right exact. Its left derived functors are the Tor modules:

$$\mathrm{Tor}_i^R(M, N) = H_i(M \otimes_R P_\bullet),$$

where $P_\bullet$ is a projective resolution of N.

An R-module M is *flat* (i.e., Tor-acyclic) if the functor $M \otimes_R -$ is exact, so that $\mathrm{Tor}_i^R(M, N) = 0$ for all $i > 0$ and every R-module N. So projective modules, and in particular free modules, are flat. In analogy with Theorem 2.23, the Tor functors can be computed using flat resolutions instead of projective resolutions.

Remark 2.34. A left exact functor $F : \mathfrak{A} \to \mathfrak{B}$ can be regarded as a right exact functor $F : \mathfrak{A}^{\mathrm{op}} \to \mathfrak{B}$. Therefore, if $\mathfrak{A}^{\mathrm{op}}$ has enough projectives, F can be derived as a right exact functor. This is what happens for the category R-**mod**, so that the contravariant functor $\mathrm{Hom}_R(-, N)$: R-**mod** $\to R$-**mod**, for a fixed R-module N, can be derived in its first argument, getting actually the same Ext functors. This generalizes what we did in Section 2.3 using free, hence projective, resolutions, and will be proved in Section 5.6.1. As a result, $\mathrm{Ext}_R^i(P, N) = 0$ for all $i > 0$ and every R-module N whenever P is a projective module. In the same way, the tensor product functor can also be derived with respect to the second argument, and therefore, $\mathrm{Tor}_i^R(M, N) = 0$ for all $i > 0$ for every R-module M when N is flat (and vice versa: if $\mathrm{Tor}_i^R(M, N) = 0$ for all $i > 0$ and every R-module M, then N is flat). One can also note that since $M \otimes_R N \simeq N \otimes_R M$ canonically, then also $\mathrm{Tor}_i^R(M, N) \simeq \mathrm{Tor}_i^R(N, M)$ for all i.

Dually to Theorem 2.17, by applying a right exact functor to an exact sequence, one obtains a long exact sequence involving the left derived functors. The proof is exactly dual to the proof of Theorem 2.17.

Theorem 2.35. *Let* $F : \mathfrak{A} \to \mathfrak{B}$ *be a right exact functor between abelian categories, and assume that* $\mathfrak{A}$ *has enough projectives. If*

$$0 \to a' \to a \to a'' \to 0$$

is an exact sequence in $\mathfrak{A}$, for every $n \geq 1$ there are morphisms δ_n : $L_nF(a'') \to L_{n-1}F(a')$ such that the sequence

$$\cdots \to L_nF(a'') \xrightarrow{\delta_n} L_{n-1}F(a') \to L_{n-1}F(a) \to L_{n-1}F(a'')$$

$$\xrightarrow{\delta_{n-1}} L_{n-2}F(a') \to \cdots \to L_1F(a'') \xrightarrow{\delta_1} F(a') \to F(a) \to F(a'') \to 0$$

is exact.

In particular, if

$$0 \to N' \to N \to N'' \to 0 \tag{2.13}$$

is an exact sequence of R-modules, and M is another R-module, there is a long exact sequence

$$\cdots \to \mathrm{Tor}_n^R(M, N'') \xrightarrow{\delta_n} \mathrm{Tor}_{n-1}^R(M, N') \to \mathrm{Tor}_{n-1}^R(M, N)$$

$$\to \mathrm{Tor}_{n-1}^R(M, N'') \xrightarrow{\delta_{n-1}} \mathrm{Tor}_{n-2}^R(M, N') \to \cdots \to \mathrm{Tor}_1^R(M, N'')$$

$$\xrightarrow{\delta_1} M \otimes_R N' \to M \otimes_R N \to M \otimes_R N'' \to 0. \tag{2.14}$$

2.8. Additional Exercises

1. (a) Define the category of divisible abelian groups.
    (b) Prove that this category is additive but not abelian.
2. An abelian group T is said to be *torsion* if all elements are torsion (see Exercise 2 in Chapter 1).

    (a) Prove that $\mathrm{Hom}_{\mathbb{Z}}(T, F) = 0$ if F is a torsion-free abelian group.
    (b) Prove that $\mathrm{Ext}_{\mathbb{Z}}^1(T, \mathbb{Z}) \simeq \mathrm{Hom}_{\mathbb{Z}}(T, \mathbb{Q}/\mathbb{Z})$ and $\mathrm{Ext}_{\mathbb{Z}}^i(T, \mathbb{Z}) = 0$ for $i > 1$.

3. Prove that $\mathbb{Q}$ is a flat a $\mathbb{Z}$-module; more generally, if D is an integral domain, and F is its field of fractions, F is a flat D-module.
4. (a) Prove that $\mathbb{Q}$ is not a projective $\mathbb{Z}$-module.
    (b) Prove that the $\mathbb{Z}$-modules $\mathbb{Z}_2$ and $\mathbb{Q}/\mathbb{Z}$ are not flat.
5. Let G be an abelian group. Show that

$$\mathrm{Ext}_{\mathbb{Z}}^1(\mathbb{Z}_n, G) \simeq G/nG \quad \text{and} \quad \mathrm{Ext}_{\mathbb{Z}}^i(\mathbb{Z}_n, G) = 0 \quad \text{for } i > 1.$$

Hint: start from the exact sequence $0 \to \mathbb{Z} \to \mathbb{Z} \to \mathbb{Z}_n \to 0$ and use Proposition 2.16.

6. Compute for every $i \geq 0$ the groups

$$\operatorname{Ext}_R^i(R/\mathfrak{m}, R/\mathfrak{m}) \quad \text{and} \quad \operatorname{Ext}_R^i(R/(y), R/(x)),$$

where $R = \Bbbk[x, y]$ is the ring of polynomials in two variables over a field $\Bbbk$, and $\mathfrak{m} = (x, y)$ is the maximal ideal of R.
Hint: use the Koszul resolution for quotients R/I [41, 64] and Exercise 20 in Chapter 1.

7. Let $\mathfrak{A}$ be an abelian category with enough injectives, fix an object $a \in \mathfrak{A}$ and assume that $\operatorname{Ext}_{\mathfrak{A}}^1(a, b) = 0$ for every object $b \in \mathfrak{A}$. Prove that $\operatorname{Hom}_{\mathfrak{A}}(a, -)$ is exact.

8. Let $\mathfrak{A}$ be an abelian category with enough injectives, and let

$$0 \to a \xrightarrow{f'} b' \to c \to 0 \quad \text{and} \quad 0 \to a \xrightarrow{f''} b'' \to c \to 0$$

be extensions in $\mathfrak{A}$ whose classes in $\operatorname{Ext}_{\mathfrak{A}}^1(c, a)$ are ξ' and ξ'', respectively. Show that the class $\xi' + \xi''$ may be represented by the extension

$$0 \to a \to b \to c \to 0,$$

where $b = b' \times_c b'' / \operatorname{im}(f', -f'')$ (Baer's sum [64]). This shows that the set $E(c, a)$ of equivalence classes of extensions of c by a, equipped with the Baer sum, is isomorphic to the group $\operatorname{Ext}_{\mathfrak{A}}^1(c, a)$.

9. Let $F : \mathfrak{A} \to \mathfrak{B}$ be a functor between abelian categories which has an exact left adjoint. Prove that F preserves injectives, i.e., $F(I)$ is an injective object of $\mathfrak{B}$ if I is an injective object of $\mathfrak{A}$.

10. (a) Prove that if M is a free R-module, any exact sequence

$$0 \to N \to P \to M \to 0$$

in R-**mod** splits.
Hint: use a basis of M to define a section of $P \to M$.
(b) Deduce that $\operatorname{Ext}_R^i(M, N) = 0$ for every $i \geq 1$ and every R-module N.

11. Use the long exact sequence (2.14) to show that if in the exact sequence (2.13) the R-modules N and N'' are flat, then N' is flat as well.
Hint: remember Remark 2.34.

12. Let S be a commutative ring, R an S-algebra with unity, and I an injective S-module. Prove that the S-module $\operatorname{Hom}_S(R, I)$ is injective. (A proof is given in [16, Lemma A3.8].)

13. Let R, S be commutative rings, and take:

- an (R, S)-bimodule M;
- an S-module N;
- an R-module P.

The Tensor-Hom adjunction [9, Section 4.1] establishes a natural isomorphism

$$\text{Hom}_S(P \otimes_R M, N) \simeq \text{Hom}_R(P, \text{Hom}_S(M, N)).$$

Use this to prove that if M is a flat R-module, the functor $\text{Hom}_R(M, -) : R\text{-}\mathbf{mod} \to R\text{-}\mathbf{mod}$ maps injective modules to injective modules.

Hint: a composition of exact functors is ...

14. Let $R\text{-}\mathbf{mod}$ be the category of modules over a ring R, and let

$$(A) \qquad 0 \to M \to N \to P \to 0,$$

where $M, N, P \in R\text{-}\mathbf{mod}$, be an extension in $R\text{-}\mathbf{mod}$ (see Section 2.4). As in Exercise 1 in Appendix A, consider a morphism $f : M \to M'$ and the morphism $\text{Ext}_R^1(P, M) \xrightarrow{f_*} \text{Ext}_R^1(P, M')$ induced by f.

Let ξ be the class in $\text{Ext}_R^1(P, M)$ of the extension (A), and let η be the class in $\text{Ext}_R^1(P, M')$ of the extension $0 \to M' \to N' \to P \to 0$ induced by f.

(a) Prove that $f_*(\xi) = \eta$.

(b) Prove that if f is an isomorphism then $N \simeq N'$.

15. Compute $\text{Tor}_i^{\mathbb{Z}}(\mathbb{Z}_2, \mathbb{Z}_2)$ for $i \geq 0$. More generally, show that

$$\text{Tor}_i^{\mathbb{Z}}(\mathbb{Z}_m, \mathbb{Z}_n) = \begin{cases} \mathbb{Z}/\gcd(m, n) & \text{for } i = 0, 1, \\ 0 & \text{for } i > 1. \end{cases}$$

16. Compute $\text{Ext}_{\mathbb{Z}}^i(\mathbb{Z}_2, \mathbb{Z}_2)$ for $i \geq 0$. More generally, show that

$$\text{Ext}_{\mathbb{Z}}^i(\mathbb{Z}_m, \mathbb{Z}_n) = \begin{cases} \mathbb{Z}/\gcd(m, n) & \text{for } i = 0, 1, \\ 0 & \text{for } i > 1. \end{cases}$$

17. Check that differential of the Chevalley–Eilenberg complex (see Example 2.31) $d : C^{\bullet}(\mathfrak{g}, M) \to C^{\bullet+1}(\mathfrak{g}, M)$ verifies $d^2 = 0$.

18. Let (ρ, M) be a representation of Lie algebra $\mathfrak{g}$ over a ring R. A *derivation* of $\mathfrak{g}$ in M is a morphism of R-modules $d : \mathfrak{g} \to M$ such that $d([x, y]) = \rho(x)(dy) - \rho(y)(dx)$. Denote by $\text{Der}(\mathfrak{g}, M)$ the R-module of derivations of $\mathfrak{g}$ in M. Every $m \in M$ defines a derivation by letting

$dx = \rho(x)(m)$; these are called *inner derivations*. Let $\text{Inn}(\mathfrak{g}, M)$ be the submodule of inner derivations.

Assume that $\mathfrak{g}$ is free over R, and prove that $\text{Der}(\mathfrak{g}, M)/\text{Inn}(\mathfrak{g}, M) \simeq H^1(\mathfrak{g}, M)$.

Hint: Use the realization of Lie algebra cohomology in terms of the Chevalley–Eilenberg complex.

19. Let $\mathfrak{k}$ be an abelian ideal in a Lie algebra $\mathfrak{g}$ and denote by $\mathfrak{q}$ the quotient Lie algebra $\mathfrak{g}/\mathfrak{k}$.

    (a) Prove that the bracket between elements of $\mathfrak{k}$ and $\mathfrak{g}$ makes $\mathfrak{k}$ into a representation of $\mathfrak{q}$.

    (b) Consider the extension (exact sequence)

    $$0 \to \mathfrak{k} \to \mathfrak{g} \to \mathfrak{q} \to 0.$$

    Prove that this defines a class in $H^2(\mathfrak{q}, \mathfrak{k})$.

20. Given a free Lie algebra $\mathfrak{g}$ over a ring R, and a representation M of $\mathfrak{g}$, the functor

    $$(-)_\mathfrak{g} : \text{Rep}(\mathfrak{g}) \to R\text{-}\mathbf{mod}$$

    $$M \mapsto M_\mathfrak{g} = M/\mathfrak{g}M$$

    is right exact, and one defines the *homology of $\mathfrak{g}$ with coefficients in M* as

    $$H_i(\mathfrak{g}, M) = L_i(M_\mathfrak{g}).$$

    (a) Show that $H_i(\mathfrak{g}, M) \simeq \text{Tor}_i^{U(\mathfrak{g})}(R, M)$ for all $i \geq 0$.

    (b) Prove that when $\mathfrak{g}$ is free over R, the R-modules $H_i(\mathfrak{g}, M)$ can be realized as the homology of the complex

    $$C_\bullet(\mathfrak{g}, M) = U(\mathfrak{g}) \otimes_R \Lambda^\bullet \mathfrak{g}$$

    with differentials

    $$\partial_i : C_n(\mathfrak{g}, M) \to C_{n-1}(\mathfrak{g}, M)$$

    given by

    $$\partial(u \otimes x_1 \wedge \cdots \wedge x_n) = \sum_{j=1}^n (-1)^{j-1} u x_j \otimes x_1 \wedge \cdots \wedge \hat{x}_j \wedge \cdots \wedge x_n$$

    $$+ \sum_{1 \leq j < k \leq n} (-1)^{j+k} u \otimes [x_j, x_k]$$

    $$\wedge x_1 \wedge \cdots \wedge \hat{x}_j \wedge \cdots \wedge \hat{x}_k \wedge \cdots \wedge x_n.$$

21. Let $\mathfrak{g}$ be a free Lie algebra over a ring R, and let M be a representation of $\mathfrak{g}$. Prove that the Lie algebra cohomology groups $H^\bullet(\mathfrak{g}, M)$ can be realized as $\mathrm{Ext}^\bullet_{U(\mathfrak{g})}(R, M)$, where M and R are regarded as a $U(\mathfrak{g})$-modules, the latter by means of the augmentation morphism $U(\mathfrak{g}) \to R$.

Chapter 3

Introduction to Sheaves

In a loose sense, a sheaf is a way of organizing local data attached to the open sets of a topological space into a global object. As such, sheaves provide a powerful tool to relate local and global features of a space, and therefore they supply a formal device to study cohomological properties of spaces. Precursors of the notion of sheaf may be found in the work of, among others, Čech and Steenrod, while the theory was first formalized by Leray in 1945. Nowadays, sheaves are pervasive in algebraic topology, algebraic and differential geometry, microlocal analysis, and more. Standard references about sheaf theory are [11, 20, 31, 59].

3.1. Presheaves and Sheaves

We shall at first introduce a more general notion, that of *presheaf*, and then, adding more structure, we shall particularize it to the notion of sheaf.

In Example 1.37 we introduced presheaves of abelian groups as functors from the category of open subsets of a topological space X to the category of abelian groups. More colloquially, a presheaf of abelian groups on X is a rule $\mathcal{P}$ which assigns an abelian group $\mathcal{P}(U)$ to each open subset U of X, and a group morphism (called restriction map) $\varphi_{U,V} : \mathcal{P}(U) \to \mathcal{P}(V)$ to each pair $V \subset U$ of open subsets, so as to verify the following requirements:

(1) $\varphi_{U,U}$ is the identity map;
(2) if $W \subset V \subset U$ are open sets, then $\varphi_{U,W} = \varphi_{V,W} \circ \varphi_{U,V}$.

The elements $s \in \mathcal{P}(U)$ are called *sections* of the presheaf $\mathcal{P}$ on U. If $s \in \mathcal{P}(U)$ is a section of $\mathcal{P}$ on U and $V \subset U$, we shall write $s_{|_V}$ instead of $\varphi_{U,V}(s)$. The restriction $\mathcal{P}_{|U}$ of $\mathcal{P}$ to an open subset U is defined in the obvious way, $\mathcal{P}_U(V) = \mathcal{P}(U \cap V)$.

Presheaves of rings are defined in the same way, by requiring that the restriction maps are ring morphisms. If $\mathcal{R}$ is a presheaf of rings on X, a presheaf $\mathcal{M}$ of abelian groups on X is called a *presheaf of modules* over $\mathcal{R}$ (or an $\mathcal{R}$-module) if, for each open subset U, $\mathcal{M}(U)$ is an $\mathcal{R}(U)$-module and for each pair $V \subset U$ the restriction map $\varphi_{U,V} : \mathcal{M}(U) \to \mathcal{M}(V)$ is a morphism of $\mathcal{R}(U)$-modules (where $\mathcal{M}(V)$ is regarded as an $\mathcal{R}(U)$-module via the restriction morphism $\mathcal{R}(U) \to \mathcal{R}(V)$). The definitions in this section are stated for the case of presheaves of abelian groups, but analogous definitions and properties hold for presheaves of rings and modules.

In the same way, a morphism of presheaves is a natural transformation between functors. This means that a morphism $f : \mathcal{P} \to \mathcal{Q}$ of presheaves over X is a family of morphisms of abelian groups $f_U : \mathcal{P}(U) \to \mathcal{Q}(U)$ for each open $U \subset X$, commuting with the restriction morphisms; i.e., the following diagram commutes:

$$
\begin{array}{ccc}
\mathcal{P}(U) & \xrightarrow{\ f_U\ } & \mathcal{Q}(U) \\
{\scriptstyle \varphi_{U,V}} \downarrow & & \downarrow {\scriptstyle \varphi_{U,V}} \\
\mathcal{P}(V) & \xrightarrow{\ f_V\ } & \mathcal{Q}(V)
\end{array}
$$

(We have denoted by the same letter the restriction morphisms of the two presheaves.)

We want to introduce the notion of *stalk* of a presheaf $\mathcal{P}$; loosely speaking, for every $x \in X$, the stalk of $\mathcal{P}$ at x is an abelian group which captures the local behaviour of $\mathcal{P}$ around x. This is done in terms of the concept of *direct limit*.

Definition 3.1. A *directed set* I is a partially ordered set such that for each pair of elements $i, j \in I$ there is a third element k such that $i < k$ and $j < k$. If I is a directed set, a directed system of abelian groups is a family $\{G_i\}_{i \in I}$ of abelian groups, such that for all $i < j$ there is a group morphism $f_{ij} : G_i \to G_j$, with $f_{ii} = id$ and $f_{jk} \circ f_{ij} = f_{ik}$ when $i < k$ and $j < k$. On the set $\mathfrak{G} = \coprod_{i \in I} G_i$, where $\coprod$ denotes disjoint union, we put the following equivalence relation: $g \sim h$, with $g \in G_i$ and $h \in G_j$, if there exists a $k \in I$ — which must satisfy $i < k$ and $j < k$ — such that

$f_{ik}(g) = f_{jk}(h)$. The direct limit $\mathfrak{l}$ of the system $\{G_i\}_{i \in I}$, denoted

$$\mathfrak{l} = \varinjlim_{i \in I} G_i,$$

is the quotient $\mathfrak{G}/\sim$.

Heuristically, two elements in $\mathfrak{G}$ represent the same element in the direct limit if they are "eventually equal". From this definition one naturally obtains the existence of canonical morphisms $G_i \to \mathfrak{l}$. The following discussion should make this notion clearer; for more detail, the reader may consult [30].

Definition 3.2. The stalk of a presheaf $\mathcal{P}$ at a point $x \in X$ is the abelian group

$$\mathcal{P}_x = \varinjlim_{U} \mathcal{P}(U),$$

where U ranges over all open neighbourhoods of x, directed by inclusion (i.e., $U < V$ if $V \subset U$).

If $x \in U$ and $s \in \mathcal{P}(U)$, the image s_x of s in $\mathcal{P}_x$ via the canonical projection $\mathcal{P}(U) \to \mathcal{P}_x$ is called the *germ* of s at x. From the very definition of direct limit we see that two sections $s \in \mathcal{P}(U)$, $s' \in \mathcal{P}(V)$, U, V being open neighbourhoods of x, define the same germ at x, i.e., $s_x = s'_x$, if and only if there exists an open neighbourhood $W \subset U \cap V$ of x such that s and s' coincide on W, $s_{|W} = s'_{|W}$.

Definition 3.3. A sheaf on a topological space X is a presheaf $\mathcal{F}$ on X which fulfils the following axioms for any open subset U of X and any open cover $\{U_i\}$ of U.

(S1) If two sections $s, \bar{s} \in \mathcal{F}(U)$ coincide when restricted to any U_i, $s_{|U_i} = \bar{s}_{|U_i}$, they are equal, $s = \bar{s}$.

(S2) Given sections $s_i \in \mathcal{F}(U_i)$ which coincide on the intersections, $s_{i|U_i \cap U_j} = s_{j|U_i \cap U_j}$ for every i, j, there exists a section $s \in \mathcal{F}(U)$ whose restriction to each U_i equals s_i, i.e., $s_{|U_i} = s_i$.

Thus, roughly speaking, sheaves are presheaves defined by local conditions. The *stalk of a sheaf* is defined as in the case of a presheaf.

Remark 3.4. A more formal definition of a sheaf can be given as follows. Let $\mathcal{P}$ be a presheaf on X, and $\{U_i\}$ a cover of an open subset $U \subset X$. If we denote by U_{ij} the intersection $U_{ij} = U_i \cap U_j$, and by $\varphi_i : \mathcal{P}(U) \to \mathcal{P}(U_i)$ and $\varphi_{ij} : \mathcal{P}(U_i) \to \mathcal{P}(U_{ij})$ the restriction morphisms, there exist a morphism

$$\mathcal{P}(U) \xrightarrow{r} \prod_i \mathcal{P}(U_i)$$

$$s \mapsto (\varphi_i(s))$$

and morphisms

$$\prod_i \mathcal{P}(U_i) \xrightarrow{r'} \prod_{ij} \mathcal{P}(U_i \cap U_j) \qquad \prod_j \mathcal{P}(U_j) \xrightarrow{r''} \prod_{ij} \mathcal{P}(U_i \cap U_j)$$

$$(s_i) \mapsto (\varphi_{ij}(s_i)) \qquad\qquad (s_j) \mapsto (\varphi_{ji}(s_j))$$

Then, axioms (S1) and (S2) are equivalent to the exactness of the sequence

$$0 \to \mathcal{P}(U) \xrightarrow{r} \prod_i \mathcal{P}(U_i) \xrightarrow{r'-r''} \prod_{ij} \mathcal{P}(U_{ij});$$

more precisely, axiom (S1) is equivalent to the injectivity of the arrow r, while (S2) is equivalent to $\ker(r' - r'') = \operatorname{im} r$.

Definition 3.5. A *separated* presheaf is a presheaf that satisfies the first sheaf axiom.

Examples of separated presheaves that are not sheaves are the constant presheaf (Example 3.9) and the presheaves of exact forms on a differentiable manifold (Example 3.13).

Example 3.6. If $\mathcal{P}$ is a separated presheaf, and $\mathcal{P}_x = \{0\}$ for all $x \in X$, then $\mathcal{P}$ is the zero presheaf, i.e., $\mathcal{P}(U) = \{0\}$ for all open sets $U \subset X$. Indeed, if $s \in \mathcal{P}(U)$, since $s_x = 0$ for all $x \in U$, there is for each $x \in U$ an open neighbourhood U_x such that $s_{|U_x} = 0$. The first sheaf axiom then implies $s = 0$. This is not true for a presheaf which is not separated, as we shall see in Example 3.22 below (this will require the concept of de Rham cohomology, which we shall sketch in Example 3.13). See also Example 3.8.

If $f : \mathcal{P} \to \mathcal{Q}$ is a morphism of presheaves on X, for every $x \in X$ there is an induced morphism between the stalks, $f_x : \mathcal{P}_x \to \mathcal{Q}_x$, defined in the following way: since the stalk $\mathcal{P}_x$ is the direct limit of the groups $\mathcal{P}(U)$ over all open U containing x, any $g \in \mathcal{P}_x$ is of the form $g = s_x$ for some open $U \ni x$ and some $s \in \mathcal{P}(U)$; then set $f_x(g) = (f_U(s))_x$.

A morphism of sheaves is just a morphism of presheaves. Sheaves of abelian groups on a given topological space make up an abelian category, as it will result from the discussion in the remaining part of this section. We shall denote this category $\mathfrak{Sh}_X$.

A sequence of morphisms of sheaves $0 \to \mathcal{F}' \to \mathcal{F} \to \mathcal{F}'' \to 0$ is *exact* if for every point $x \in X$, the sequence of morphisms between the stalks

$$0 \to \mathcal{F}'_x \to \mathcal{F}_x \to \mathcal{F}''_x \to 0$$

is exact.

Remark 3.7. If $0 \to \mathcal{F}' \to \mathcal{F} \to \mathcal{F}'' \to 0$ is an exact sequence of sheaves, for every open subset $U \subset X$ the sequence of groups $0 \to \mathcal{F}'(U) \to \mathcal{F}(U) \to \mathcal{F}''(U)$ is exact, but the last arrow may fail to be surjective. Instances of this situation are shown in Examples 3.13 and 3.14.

Example 3.8. Let $f : \mathcal{F} \to \mathcal{G}$ be a surjective sheaf morphism which is not surjective as a presheaf morphism. Then the presheaf

$$(\operatorname{coker} f)(U) = \mathcal{G}(U)/f_U(\mathcal{F}(U))$$

is not separated (see Examples 3.13 and 3.14).

Example 3.9. Let G be an abelian group. Defining $\mathcal{P}(U) \equiv G$ for every open subset U and taking the identity maps as restriction morphisms, we obtain a presheaf, called the *constant presheaf* $\tilde{G}_X$. All stalks $(\tilde{G}_X)_x$ of $\tilde{G}_X$ are isomorphic to the group G. The presheaf $\tilde{G}_X$ in general is not a sheaf: if V_1 and V_2 are disjoint open subsets of X, and $U = V_1 \cup V_2$, the sections $g_1 \in \tilde{G}_X(V_1) = G$, $g_2 \in \tilde{G}_X(V_2) = G$, with $g_1 \neq g_2$, satisfy the hypothesis of the second sheaf axiom (S2) (since $V_1 \cap V_2 = \emptyset$ there is nothing to satisfy), but there is no section $g \in \tilde{G}_X(U) = G$ which restricts to g_1 on V_1 and to g_2 on V_2.

Example 3.10. Let $\mathcal{C}_X(U)$ be the ring of real-valued continuous functions on an open set U of X. Then $\mathcal{C}_X$ is a sheaf (with the obvious restriction morphisms), the sheaf of continuous functions on X. The stalk $\mathcal{C}_x \equiv (\mathcal{C}_X)_x$ at x is the ring of germs of continuous functions at x.

Example 3.11. In the same way one can define the following sheaves:

- Let X be a differentiable manifold; then one may consider the sheaf $\mathcal{C}_X^\infty$ of differentiable functions on X; the sheaf of vector fields on X, also called the tangent sheaf to X; the sheaves Ω_X^p of differential p-forms, and all the sheaves of tensor fields on X.

- Given a complex manifolds X one can consider on it the sheaf of holomorphic functions, the sheaves of holomorphic p-forms, and the sheaves of differential forms of type (p, q).

These examples are treated in many standard texts in real and complex geometry, see, e.g., [7, 21, 39, 58, 63, 65].

Example 3.12. If X is a scheme (see Section 4.5), one can introduce the sheaf of Kähler differentials on X; its pth exterior power is the sheaf of regular p-forms on X [28].

Example 3.13. Let X be a differentiable manifold, and let $d : \Omega_X^\bullet \to \Omega_X^{\bullet+1}$ be the exterior differential. We can define the presheaves $\mathcal{Z}_X^p$ of closed differential p-forms, and $\mathcal{B}_X^p$ of exact differential p-forms,

$$\mathcal{Z}_X^p(U) = \{\omega \in \Omega_X^p(U) \mid d\omega = 0\},$$

$$\mathcal{B}_X^p(U) = \{\omega \in \Omega_X^p(U) \mid \omega = d\tau \quad \text{for some } \tau \in \Omega_X^{p-1}(U)\}.$$

The *pth de Rham cohomology group of U* (and so, in particular of X) is by definition the quotient

$$H_{dR}^p(U) = \mathcal{Z}_X^p(U)/\mathcal{B}_X^p(U),$$

i.e., it is the pth cohomology of the de Rham complex of differential forms $(\Omega^\bullet(U), d)$.

Now $\mathcal{Z}_X^p$ is a sheaf, since the condition of being closed is local: a differential form is closed if and only if it is closed in a neighbourhood of each point of X. On the contrary, $\mathcal{B}_X^p$ is not a sheaf. In fact, if $X = \mathbb{R}^2$, the presheaf $\mathcal{B}_X^1$ of exact differential 1-forms does not fulfil the second sheaf axiom: consider the form

$$\omega = \frac{xdy - ydx}{x^2 + y^2}$$

defined on the open subset $U = X - \{(0, 0)\}$. Since ω is closed on U, there is an open cover $\{U_i\}$ of U by open subsets where ω is an exact form, $\omega|_{U_i} \in \mathcal{B}_X^1(U_i)$ (this is Poincaré Lemma, see, e.g., [7, 58, 63]). But ω is not an exact form on U because its integral along the unit circle γ is different from 0:

$$\int_\gamma \omega = \int_\gamma \left[\frac{x}{x^2 + y^2}dy - \frac{y}{x^2 + y^2}dx\right] = 2\pi$$

(we have chosen the counterclockwise orientation of γ).

This means that, while the sequence of sheaf morphisms

$$0 \to \mathbb{R} \to \mathcal{C}_X^\infty \xrightarrow{d} \mathcal{Z}_X^1 \to 0$$

is exact, the morphism $\mathcal{C}_X^\infty(U) \xrightarrow{d} \mathcal{Z}_X^1(U)$ may fail to be surjective.

Example 3.14. Let X be a complex manifold, $\mathbb{Z}_X$ the (locally) constant sheaf (i.e., the sheaf associated with the constant presheaf, see Example 3.9), $\mathcal{O}_X$ the sheaf of holomorphic functions on X, and $\mathcal{O}_X^*$ the sheaf of nowhere vanishing holomorphic functions. In analogy with the exact sequence in Example 1.16, we may consider the sequence

$$0 \to \mathbb{Z}_X \to \mathcal{O}_X \xrightarrow{\exp} \mathcal{O}_X^* \to 1. \tag{3.1}$$

This is an exact sequence of sheaves, in particular $\exp : \mathcal{O}_X \to \mathcal{O}_X^*$ is surjective as a map of sheaves, since in a neighbourhood of every point, an inverse may be found by applying the logarithm function. However, since the latter is multi-valued, surjectivity may fail on nonsimply connected open sets. See Example 4.35.

3.2. Étalé Space

We wish now to describe how, given a presheaf, one can naturally associate with it a sheaf having the same stalks. This is accomplished by associating to a given presheaf $\mathcal{P}$ on a topological space X another topological space, fibred over X, called the *étalé space* of the presheaf $\mathcal{P}$, and taking the sheaf of its continuous sections. Actually, this notion of étalé space turns out to be useful also in several other situations.

As a first step we consider the case of a constant presheaf $\tilde{G}_X$ on a topological space X, where G is an abelian group. We can define another presheaf G_X on X by putting $G_X(U) = \{$locally constant functions $f : U \to G\}$,[a] where $\tilde{G}_X(U) = G$ is included as the constant functions. It is quite clear that G_X is a sheaf, called the *locally constant sheaf with stalk G*. Note that the functions $f : U \to G$ are sections of the projection $\pi : \coprod_{x \in X}(G_X)_x \to X$, and the locally constant functions correspond to those sections which locally coincide with the sections produced by the elements of G.

Now, let $\mathcal{P}$ be an arbitrary presheaf on X. Consider the disjoint union of the stalks $\underline{\mathcal{P}} = \coprod_{x \in X} \mathcal{P}_x$ and the natural projection $\pi : \underline{\mathcal{P}} \to X$.

---

[a]A function is locally constant on U if every point in U has a neighbourhood over which f is constant.

The sections $s \in \mathcal{P}(U)$ of the presheaf $\mathcal{P}$ on an open subset U produce sections $\underline{s} : U \hookrightarrow \underline{\mathcal{P}}$ of π, defined by $\underline{s}(x) = s_x$, and we can define a new presheaf $\mathcal{P}^\natural$ by taking $\mathcal{P}^\natural(U)$ as the group of those sections $\sigma : U \hookrightarrow \underline{\mathcal{P}}$ of π such that for every point $x \in U$ there is an open neighbourhood $V \subset U$ of x which satisfies $\sigma_{|V} = \underline{s}$ for some $s \in \mathcal{P}(V)$. That is, $\mathcal{P}^\natural$ is the presheaf of all sections that locally coincide with sections of $\mathcal{P}$. It can be described in another way by the following construction.

Definition 3.15. The set $\underline{\mathcal{P}}$, endowed with the topology whose base of open subsets consists of the sets $\underline{s}(U)$, for U open in X and $s \in \mathcal{P}(U)$, is called the étalé space of the presheaf $\mathcal{P}$. The sheaf $\mathcal{P}^\natural$ associated to $\mathcal{P}$ is the sheaf of continuous sections of $\pi : \underline{\mathcal{P}} \to X$.

Note that by construction there is a natural morphism $\mathcal{P} \to \mathcal{P}^\natural$, which is an isomorphism on the stalks. As a consequence of the sheaf axioms, this is an isomorphism if and only if $\mathcal{P}$ is a sheaf.

Lemma 3.16. *If* $f : \mathcal{P} \to \mathcal{Q}$ *is a morphism of presheaves on* X, *there is a unique morphism* $f^\natural : \mathcal{P}^\natural \to \mathcal{Q}^\natural$ *such that the diagram*

$$
\begin{array}{ccc}
\mathcal{P} & \xrightarrow{\ f\ } & \mathcal{Q} \\
\downarrow & & \downarrow \\
\mathcal{P}^\natural & \xrightarrow{\ f^\natural\ } & \mathcal{Q}^\natural
\end{array}
$$

commutes.

Proof. For every $x \in X$ the morphism f induces a morphism between the stalks $f_x : \mathcal{P}_x \to \mathcal{Q}_x$. These in turn induce a morphism between the étalé spaces $\underline{f} : \underline{\mathcal{P}} \to \underline{\mathcal{Q}}$. Now $f^\natural : \mathcal{P}^\natural \to \mathcal{Q}^\natural$ is the morphism induced by letting

$$\underline{f}^\natural(s)(x) = \underline{f}(\underline{s}(x))$$

for every $s \in \mathcal{P}^\natural(U)$. To prove the uniqueness note that if g is another such morphism, for every $x \in X$ one has $g_x = f_x = (f^\natural)_x$, which by the sheaf axioms implies $g = f^\natural$. $\qquad\square$

Proposition 3.17. *The sheaf* $\mathcal{P}^\natural$ *defined above enjoys the following universal property: if* $\mathcal{F}$ *is a sheaf, every morphism* $f : \mathcal{P} \to \mathcal{F}$ *factors uniquely through* $\mathcal{P}^\natural$; *that is, there is a unique morphism* $f^\natural : \mathcal{P}^\natural \to \mathcal{F}$ *such that the* *diagram*

$$
\begin{array}{ccc}
\mathcal{P} & \xrightarrow{\ f\ } & \mathcal{F} \\
\downarrow & \nearrow{\scriptstyle f^\natural} & \\
\mathcal{P}^\natural & &
\end{array}
$$

commutes.

Proof. Apply Lemma 3.16 with $\mathcal{Q} = \mathcal{F}$, and note that the right vertical arrow there is an isomorphism. $\qquad\square$

Exercise 3.18. Show that $\pi : \mathcal{P} \to X$ is a local homeomorphism, i.e., every point $u \in \mathcal{P}$ has an open neighbourhood U such that $\pi : U \to \pi(U)$ is a homeomorphism.

Exercise 3.19. Show that for every open set $U \subset X$ and every $s \in \mathcal{P}(U)$, the section $\underline{s} : U \to \mathcal{P}$ is continuous.

Example 3.20. The morphism $\phi : \mathcal{P} \to \mathcal{P}^\natural$ may be neither injective nor surjective: for instance, the morphism between the constant presheaf $\tilde{G}_X$ and its associated sheaf G_X in general is injective but nor surjective.

Example 3.21. Let $0 \to \mathcal{F}' \to \mathcal{F} \to \mathcal{F}'' \to 0$ be an exact sequence of sheaves, and let $\mathcal{Q}$ be the quotient presheaf $\mathcal{F}/\mathcal{F}'$, i.e., for every open set $U \subset X$,

$$
\mathcal{Q}(U) = \frac{\mathcal{F}(U)}{\mathcal{F}'(U)}.
$$

Then one can show that the quotient sheaf $\mathcal{F}''$ is isomorphic to the "sheafification" $\mathcal{Q}^\natural$ of the quotient presheaf. Moreover, in this case the natural morphism $\mathcal{Q} \to \mathcal{Q}^\natural$ is injective,[b] as one can see by applying the Snake Lemma (see Appendix A.2) to the diagram

$$
\begin{array}{ccccccccc}
0 & \longrightarrow & \mathcal{F}'(U) & \longrightarrow & \mathcal{F}(U) & \longrightarrow & \mathcal{Q}(U) & \longrightarrow & 0 \\
 & & \| & & \| & & \downarrow & & \\
0 & \longrightarrow & \mathcal{F}'(U) & \longrightarrow & \mathcal{F}(U) & \longrightarrow & \mathcal{Q}^\natural(U) & &
\end{array}
$$

Note that $\mathcal{Q}^\natural \simeq \mathcal{F}''$.

---

[b]That is, $\mathcal{Q}$ is a separated presheaf, see Definition 3.5 and Exercise 4 in this chapter.

Example 3.22. As a second example we study the sheaf associated with the presheaf $\mathcal{B}_X^p$ of exact p-forms on a differentiable manifold X. For any open set U we have an exact sequence of abelian groups (actually of $\mathbb{R}$-vector spaces)

$$0 \to \mathcal{B}_X^p(U) \to \mathcal{Z}_X^p(U) \to \mathcal{H}_X^p(U) \to 0,$$

where $\mathcal{H}_X^p$ is the presheaf that with any open set U associates its pth de Rham cohomology group, $\mathcal{H}_X^p(U) = H_{dR}^p(U)$. Now, the open neighbourhoods of any point $x \in X$ which are diffeomorphic to $\mathbb{R}^n$ (where $n = \dim X$) are cofinal[c] in the family of all open neighbourhoods of x, so that $(\mathcal{H}_X^p)_x = 0$ by the Poincaré Lemma. In accordance with Example 3.6 this means that $(\mathcal{H}_X^p)^\natural = 0$, which is tantamount to $(\mathcal{B}_X^p)^\natural \simeq \mathcal{Z}_X^p$.

In this case, the natural morphism $\mathcal{H}_X^p \to (\mathcal{H}_X^p)^\natural$ is of course surjective but not injective, while $\mathcal{B}_X^p \to (\mathcal{B}_X^p)^\natural = \mathcal{Z}_X^p$ is injective but not surjective.

The étalé space of a sheaf also allows one to define sections of a sheaf on any subset of X, whether it is open or not.

Definition 3.23. Given a sheaf $\mathcal{F}$ on a topological space X and a subset (not necessarily open) $S \subset X$, the sections of the sheaf $\mathcal{F}$ on S are the continuous sections $\sigma : S \hookrightarrow \underline{F}$ of $\pi : \underline{F} \to X$. The group of such sections is denoted by $\Gamma(S, \mathcal{F})$.

3.3. Some Constructions

In this section, we introduce the notions of direct sum of presheaves and sheaves, Hom groups and sheaves and the tensor product operation.

Definition 3.24. Let $\mathcal{P}$, $\mathcal{Q}$ be presheaves on a topological space X.[d]

(1) The direct sum of $\mathcal{P}$ and $\mathcal{Q}$ is the presheaf $\mathcal{P} \oplus \mathcal{Q}$ given, for every open subset $U \subset X$, by $(\mathcal{P} \oplus \mathcal{Q})(U) = \mathcal{P}(U) \oplus \mathcal{Q}(U)$ with the obvious restriction morphisms.

---

[c]Let I be a directed set. A subset J of I is said to be *cofinal* if for any $i \in I$ there is a $j \in J$ such that $i < j$. By the definition of direct limit we see that, given a directed family of abelian groups $\{G_i\}_{i \in I}$, if $\{G_j\}_{j \in J}$ is the subfamily indexed by J, then

$$\varinjlim_{i \in I} G_i \simeq \varinjlim_{j \in J} G_j;$$

that is, direct limits can be taken over cofinal subsets of the index set.

[d]Since we are dealing with abelian groups, i.e. with $\mathbb{Z}$-modules, the $\mathcal{H}om$ modules and tensor products are taken over $\mathbb{Z}$.

(2) Hom($\mathcal{P}, \mathcal{Q}$) is the space of morphisms from $\mathcal{P}$ to $\mathcal{Q}$; this is an abelian group in a natural manner. The presheaf of homomorphisms is the presheaf $\mathcal{H}om(\mathcal{P}, \mathcal{Q})$ given by $\mathcal{H}om(\mathcal{P}, \mathcal{Q})(U) = \text{Hom}(\mathcal{P}_{|U}, \mathcal{Q}_{|U})$ with the natural restriction morphisms.

(3) The tensor product of $\mathcal{P}$ and $\mathcal{Q}$ is the presheaf $(\mathcal{P} \otimes \mathcal{Q})(U) = \mathcal{P}(U) \otimes \mathcal{Q}(U)$.

Note that $\text{Hom}(\mathcal{P}, \mathcal{Q}) = \Gamma(X, \mathcal{H}om(\mathcal{P}, \mathcal{Q}))$ by definition.

If $\mathcal{F}$ and $\mathcal{G}$ are sheaves, then the presheaves $\mathcal{F} \oplus \mathcal{G}$ and $\mathcal{H}om(\mathcal{F}, \mathcal{G})$ are sheaves. On the contrary, the tensor product of $\mathcal{F}$ and $\mathcal{G}$ previously defined may not be a sheaf, and one defines the tensor product of the sheaves $\mathcal{F}$ and $\mathcal{G}$ as the sheaf associated with the presheaf $U \to \mathcal{F}(U) \otimes \mathcal{G}(U)$.

It should be noticed that $\mathcal{H}om(\mathcal{F}, \mathcal{G})(U) \not\simeq \text{Hom}(\mathcal{F}(U), \mathcal{G}(U))$ in general, and $\mathcal{H}om(\mathcal{F}, \mathcal{G})_x \not\simeq \text{Hom}(\mathcal{F}_x, \mathcal{G}_x)$. We refer the reader to [28, Proposition III.6.8] or [49, 17.20] for conditions ensuring that $\mathcal{H}om(\mathcal{F}, \mathcal{G})_x \simeq \text{Hom}(\mathcal{F}_x, \mathcal{G}_x)$.

Example 3.25. There is a natural morphism

$$\mathcal{H}om(\mathcal{F}, \mathcal{G})_x \to \text{Hom}(\mathcal{F}_x, \mathcal{G}_x),$$

but this may fail to be an isomorphism. Let R be a commutative ring with unity, and let $X = \text{Spec}\, R$ be its spectrum [28] (a notion that we shall explain in Section 4.5). Choose $\mathcal{F} = \bigoplus_{i \in I} \mathcal{O}_X$, where I is an infinite set, and $\mathcal{O}_X$ is the structure sheaf of X. Also, choose $\mathcal{G} = \mathcal{O}_X$. Then,

$$\mathcal{H}om_{\mathcal{O}_X}(\mathcal{F}, \mathcal{G})_x \simeq \left[\mathcal{H}om_{\mathcal{O}_X}\left(\bigoplus_{i \in I} \mathcal{O}_X, \mathcal{O}_X\right)\right]_x \simeq \left(\prod_{i \in I} \mathcal{O}_X\right)_x,$$

$$\text{Hom}(\mathcal{F}_x, \mathcal{G}_x) \simeq \text{Hom}_{\mathcal{O}_{X,x}}\left(\bigoplus_{i \in I} \mathcal{O}_{X,x}, \mathcal{O}_{X,x}\right) \simeq \prod_{i \in I} \mathcal{O}_{X,x};$$

the two groups are in general different as infinite products do not commute with taking direct limits [5] (which are involved in the definition of stalk). An example is provided by $R = \mathbb{Z}$ (see Example 4.52).

Example 3.26. An example of tensor product of sheaves which is different from the tensor product taken as presheaves is the following. Take $X = \mathbb{P}^1$, the complex projective line (see p. 110), and let $\mathcal{O}_{\mathbb{P}^1}(-1)$ be the tautological line bundle; the dual line bundle is denoted by $\mathcal{O}_{\mathbb{P}^1}(1)$ [21, 28].

The tautological line bundle $\mathcal{O}_{\mathbb{P}^1}(-1)$ has no global sections, so that

$$\mathcal{O}_{\mathbb{P}^1}(-1)(\mathbb{P}^1) \otimes_{\mathcal{O}_{\mathbb{P}^1}(\mathbb{P}^1)} \mathcal{O}_{\mathbb{P}^1}(1)(\mathbb{P}^1) = 0.$$

On the other hand, the tensor product sheaf is

$$\mathcal{O}_{\mathbb{P}^1}(-1) \otimes_{\mathcal{O}_{\mathbb{P}^1}} \mathcal{O}_{\mathbb{P}^1}(1) \simeq \mathcal{O}_{\mathbb{P}^1}$$

whose space of global sections is $\mathbb{C}$. Note that in this example the natural morphism $\mathcal{P} \to \mathcal{P}^\natural$, where $\mathcal{P}$ is the tensor product of $\mathcal{O}_{\mathbb{P}^1}(-1)$ and $\mathcal{O}_{\mathbb{P}^1}(1)$ as presheaves, is not surjective.

Example 3.27. It may happen that tensor product presheaf and sheaf coincide. Let $\mathcal{P}$ be the tensor product $\mathbb{Z}_X \otimes \mathbb{Z}_X$ as presheaves, where $\mathbb{Z}_X$ is the constant sheaf with stalk $\mathbb{Z}$, and let U be an open subset of X. Then

$$\mathcal{P}(U) = \mathbb{Z}_X(U) \otimes_{\mathbb{Z}_X(U)} \mathbb{Z}_X(U) \simeq \mathbb{Z}_X(U) \simeq (\mathbb{Z}_X \otimes_{\mathbb{Z}_X} \mathbb{Z}_X)(U).$$

3.4. Direct and Inverse Images

Here, we study the behaviour of presheaves and sheaves under change of base space. Let $f : X \to Y$ be a continuous map. We start by considering *direct images*.

Definition 3.28. The direct image by f of a presheaf $\mathcal{P}$ on X is the presheaf $f_*\mathcal{P}$ on Y defined by $(f_*\mathcal{P})(V) = \mathcal{P}(f^{-1}(V))$ for every open subset $V \subset Y$.

If $\mathcal{F}$ is a sheaf on X, then $f_*\mathcal{F}$ turns out to be a sheaf. This defines an additive functor $f_* : \mathfrak{Sh}_X \to \mathfrak{Sh}_Y$ between the categories of sheaves of abelian groups on X and Y. By Remark 3.7 this functor is left exact.

Now we consider inverse images. Let $\mathcal{P}$ be a presheaf on Y.

Definition 3.29. The inverse image of $\mathcal{P}$ by f is the presheaf on X defined by

$$U \rightsquigarrow \varinjlim_{U \subset f^{-1}(V)} \mathcal{P}(V).$$

The inverse image sheaf of a sheaf $\mathcal{F}$ on Y is the sheaf $f^{-1}\mathcal{F}$ associated with the inverse image presheaf of $\mathcal{F}$.

The stalk of the inverse image presheaf at a point $x \in X$ is isomorphic to $\mathcal{P}_{f(x)}$ (cf. Exercise 19). It follows that if $0 \to \mathcal{F}' \to \mathcal{F} \to \mathcal{F}'' \to 0$ is an exact sequence of sheaves on Y, the induced sequence

$$0 \to f^{-1}\mathcal{F}' \to f^{-1}\mathcal{F} \to f^{-1}\mathcal{F}'' \to 0$$

of sheaves on X is also exact (that is, the inverse image functor for sheaves of abelian groups is exact).

The inverse image sheaf can also be described using the notion of fibred product. Given topological spaces X, Y, Z, with continuous maps $f : X \to Y$, $g : Z \to Y$, the fibred product $X \times_Y Z$ is the subspace of the cartesian product $X \times Z$ formed by pairs (x, z) such that $f(x) = g(z)$. It is topologized with the relative topology. The natural projections p_1, p_2 onto the factors of the cartesian product restrict to the fibred product, and one has a commutative diagram

$$
\begin{array}{ccc}
X \times_Y Z & \xrightarrow{\ p_2\ } & Z \\
\downarrow{\scriptstyle p_1} & & \downarrow{\scriptstyle g} \\
X & \xrightarrow{\ f\ } & Y
\end{array}
$$

For properties of the fibred product of topological spaces the reader can consult [12]. Here, we only cite its *universal property*: if T is a topological space with maps $T \to X$, $T \to Z$ such that the diagram formed by the outer maps in

$$
\begin{array}{ccc}
T & & \\
 & X \times_Y Z \longrightarrow Z & \\
 & \downarrow \qquad\qquad \downarrow & \\
 & X \longrightarrow Y &
\end{array}
$$

commutes, then there is a unique map $T \to X \times_Y Z$ (the dotted arrow in the previous diagram) such that the full diagram commutes.

Remark 3.30. Actually the notion of fibred product (also called pullback) exists in any category, see Appendix A.1.

Exercise 3.31. Show that the étalé space of the inverse image sheaf $f^{-1}\mathcal{G}$ of a sheaf $\mathcal{G}$ on Y can be identified with the fibred product $X \times_Y \underline{\mathcal{G}}$. Thus the étalé space of $f^{-1}\mathcal{G}$ fits into a cartesian diagram (i.e., a fibred product diagram)

$$
\begin{array}{ccc}
f^{-1}\mathcal{G} & \overset{f}{\longrightarrow} & \mathcal{G} \\
\downarrow & & \downarrow \\
X & \overset{f}{\longrightarrow} & Y
\end{array}
$$

Exercise 3.32. Show that the inverse image of the constant sheaf G_Y on Y with stalk G is the constant sheaf G_X with stalk G, $f^{-1}G_Y \simeq G_X$ (this is easily shown in terms of the characterization via fibred product of Exercise 3.31).

Using the language introduced in Section 1.1.2, one can see that $f_* : \mathfrak{Sh}_X \to \mathfrak{Sh}_Y$ and $f^{-1} : \mathfrak{Sh}_Y \to \mathfrak{Sh}_X$ are a pair of *adjoint functors*. Indeed, given sheaves $\mathcal{F}$ on X and $\mathcal{G}$ on Y, one can define natural morphisms

$$
\phi : \mathcal{G} \to f_* f^{-1}\mathcal{G}, \quad \psi : f^{-1}f_*\mathcal{F} \to \mathcal{F}.
$$

For instance, the first morphism is defined by the following construction: if $U \subset Y$ is an open subset, and $s \in \mathcal{G}(U)$, we build the diagram

$$
\begin{array}{ccc}
f^{-1}(U) & & \\
& \searrow^{\,s \circ f} & \\
& f^{-1}\mathcal{G} \longrightarrow \mathcal{G} & \\
& \downarrow \qquad\quad \downarrow & \\
& X \overset{f}{\longrightarrow} Y &
\end{array}
$$

where the leftmost arrow is the inclusion. Then the dotted arrow, which exists by the universal property of the fibred product, defines the section $\phi(s) \in f^{-1}\mathcal{G}(f^{-1}(U)) = (f_* f^{-1}\mathcal{G})(U)$.

The morphism ψ is defined by the composition shown in the diagram

$$
\begin{array}{ccc}
V & & \\
\downarrow{\scriptstyle t} & \searrow^{\,\psi(t)} & \\
f^{-1}f_*\mathcal{F} & \longrightarrow & f_*\mathcal{F} \\
\downarrow & & \downarrow \\
X & \longrightarrow & Y
\end{array}
$$

where V is an open subset of X and $t \in (f^{-1}f_*\mathcal{F})(V)$.

The morphisms ϕ and ψ induce morphisms

$$\Phi : \mathrm{Hom}(f^{-1}\mathcal{G}, \mathcal{F}) \to \mathrm{Hom}(\mathcal{G}, f_*\mathcal{F}), \quad \Psi : \mathrm{Hom}(\mathcal{G}, f_*\mathcal{F}) \to \mathrm{Hom}(f^{-1}\mathcal{G}, \mathcal{F})$$

by letting

$$\Phi(g) = f_*(g) \circ \phi, \quad \Psi(h) = \psi \circ f^{-1}(h)$$

if $g \in \mathrm{Hom}(f^{-1}\mathcal{G}, \mathcal{F})$, $h \in \mathrm{Hom}(\mathcal{G}, f_*\mathcal{F})$. These morphisms are one the inverse of the other and establish the adjunction

$$\mathrm{Hom}(f^{-1}\mathcal{G}, \mathcal{F}) \simeq \mathrm{Hom}(\mathcal{G}, f_*\mathcal{F}). \tag{3.2}$$

3.5. Additional Exercises

1. Prove that if $\mathcal{P}$ is a separated presheaf (see Definition 3.5) then $\mathcal{P}(\emptyset) = 0$.[e]

2. (a) Let $\mathcal{P}$ be a separated presheaf on a topological space X. Prove that the natural morphism $\mathcal{P}(X) \to \prod_{x \in X} \mathcal{P}_x$ is injective. Find a presheaf for which this is not true.

   (b) Let $f : \mathcal{P} \to \mathcal{Q}$ be a presheaf morphism, where $\mathcal{Q}$ is separated. Assume that for every $x \in X$ the morphism $f_x : \mathcal{P}_x \to \mathcal{Q}_x$ is zero; prove that $f = 0$.

3. Let $0 \to \mathcal{F}' \to \mathcal{F} \to \mathcal{F}'' \to 0$ be an exact sequence of sheaves. Show that $0 \to \mathcal{F}' \to \mathcal{F} \to \mathcal{F}''$ is an exact sequence of presheaves.

4. Prove that a presheaf $\mathcal{P}$ is separated if and only if the morphism $\mathcal{P} \to \mathcal{P}^\natural$ is injective.

5. Let $\mathbb{R}$ be the real line with the standard Euclidean topology. Let $\mathcal{B}$ be the presheaf

$$\mathcal{B}(U) = \{\text{continuous bounded functions } U \to \mathbb{R}\}.$$

   (a) Prove that $\mathcal{B}$ is a separated presheaf.

   (b) Prove that $\mathcal{B}$ is not a sheaf.

   (c) Determine the associated sheaf $\mathcal{B}^\natural$.

6. Check that the real-valued functions on a topological space make up a sheaf.

7. (a) Prove that the support of a section of a sheaf of abelian groups is a closed subset.

---

[e]This requires to use the fact that a direct product over an empty index set is the terminal object in the category.

(b) Give an example of a sheaf of abelian groups whose support is not closed.

8. Given a sheaf morphism $f : \mathcal{F} \to \mathcal{G}$, show that for any $x \in X$, the stalk at x of the sheaf $\ker f$ is isomorphic to $\ker f_x : \mathcal{F}_x \to \mathcal{G}_x$. Analogously, show that $(\operatorname{coker} f)_x \simeq \operatorname{coker} f_x$.

9. Prove that a sheaf morphism is an isomorphism if and only if it is both injective and surjective (where surjective means that it is surjective on every stalk).

10. Let $\mathcal{P}$ and $\mathcal{F}$ be a presheaf and a sheaf on a topological space X, respectively. Let $f : \mathcal{P} \to \mathcal{F}$ be a morphism, and assume that

$$f_x : \mathcal{P}_x \to \mathcal{F}_x$$

is an isomorphism for every $x \in X$. Prove that $\mathcal{F} \simeq \mathcal{P}^\natural$.

11. (a) If $0 \to \mathcal{F}' \to \mathcal{F} \to \mathcal{F}'' \to 0$ is an exact sequence of sheaves, show that $\mathcal{F}''$ is the sheaf associated to the presheaf $U \rightsquigarrow \mathcal{Q}(U) = \mathcal{F}(U)/\mathcal{F}'(U)$.

    (b) Show an example where $\mathcal{Q}$ is not a sheaf.

12. Prove that the cokernel of a sheaf morphism, defined as the sheaf associated with the cokernel presheaf, satisfies the universal property of cokernels (see footnote b in Chapter 1).

13. A nonempty topological space X is *irreducible* if it cannot be written as the union of two proper closed subsets.[f] Let X be an irreducible topological space. Show that, given any abelian group G, the constant presheaf G_X is a sheaf.

14. Prove that the locally constant sheaf associated with an abelian group G and the constant sheaf are isomorphic.

15. Let $\mathcal{R}$ be a presheaf of commutative rings with unity on a topological space X, and let $\mathcal{R}^*$ be the presheaf of invertible sections of $\mathcal{R}$. Fix a global section s of $\mathcal{R}$. Define the subset S of X

$$S = \{x \in X \mid s_x \in \mathcal{R}_x^*\}.$$

Prove that S is open.

16. Let $f : X \to Y$ be a continuous map of topological spaces.

    (a) Let $\mathcal{F}$ be a sheaf of abelian groups on X; prove that $f_*\mathcal{F}$ is a sheaf.

    (b) Let G_X be a constant sheaf on X. Prove that f_*G_X need not be a constant sheaf.

---

[f]Examples of irreducible topological spaces are easily exhibited considering the *Zariski topology*, which we will introduce later on (see Section 4.5).

(c) We know that if $0 \to \mathcal{F}' \to \mathcal{F} \to \mathcal{F}'' \to 0$ is an exact sequence of sheaves on X, the sequence of sheaves on Y

$$0 \to f_*\mathcal{F}' \to f_*\mathcal{F} \to f_*\mathcal{F}'' \to 0$$

need not be exact. Provide an example of such a situation.

17. Let $X = \mathbb{C}$ with the Euclidean topology, and consider the sheaf morphism $\exp : \mathcal{O}_X \to \mathcal{O}_X^*$, where $\mathcal{O}_X$ is the sheaf of holomorphic functions, and $\mathcal{O}_X^*$ is the sheaf of nowhere vanishing holomorphic functions; $\exp$ is the morphism $\exp(f) = e^{2\pi i f}$. Find an open set on which $\exp$ is not surjective.

18. For a topological space X denote by $\mathfrak{Psh}_X$ the category of presheaves of abelian groups on X.

   (a) Prove that the forgetful functor $FF : \mathfrak{Sh}_X \to \mathfrak{Psh}_X$ is additive and left exact.
   (b) Prove that the "sheafification" functor $\natural : \mathfrak{Psh}_X \to \mathfrak{Sh}_X$ is exact and is left adjoint to FF. (Note that this provides an example of a functor as in Exercise 9 of Chapter 2.)

19. Let X be a topological space, $Z \subset X$ a closed subset considered as a topological space with the relative topology, $\mathcal{F}$ a sheaf of abelian groups on Z.

   (a) Prove, for every $x \in Z$, the isomorphism of stalks $(i_*\mathcal{F})_x \simeq \mathcal{F}_x$, where $i : Z \to X$ is the inclusion.
   (b) Deduce that the functor $i_* : \mathfrak{Sh}_Z \to \mathfrak{Sh}_X$ is exact.
   (c) Let $\mathcal{G}$ be a sheaf of abelian groups on X. Prove that the sequence

   $$0 \to j_!(\mathcal{G}_{|X-Z}) \to \mathcal{G} \to i_*(\mathcal{G}_{|Z}) \to 0$$

   is exact, where $j : X - Z \to X$ is the inclusion, and $j_!$ is the functor defined in equation (4.9). Given the pair $(X, \mathcal{G})$, this is the exact sequence associated to a closed subset of X.

20. (a) Given a continuous map of topological spaces $f : X \to Y$, and a sheaf of abelian groups $\mathcal{F}$ on Y, prove that for every $x \in X$ the stalk $(f^{-1}\mathcal{F})_x$ is isomorphic to $\mathcal{F}_{f(x)}$.
   (b) Check that the inverse image of a constant sheaf on Y is a constant sheaf with the same stalks.

21. For any topological space X involved, we shall denote by G_X the constant sheaf on X with stalk an abelian group G. Let Y be a

topological space with two irreducible[g] components Y_1 and Y_2, and let $Z = Y_1 \cap Y_2$ be nonempty. Denote by $i_1 : Y_1 \to Y$, $i_2 : Y_2 \to Y$, $i_Z : Z \to Y$ the inclusions. Prove that there is an exact sequence (Mayer–Vietoris sequence)

$$0 \to G_Y \to i_{1*}G_{Y_1} \oplus i_{2*}G_{Y_2} \to i_{Z*}G_Z \to 0. \qquad (3.3)$$

22. Let X be an irreducible topological space. Prove that

$$\mathrm{Hom}(\mathbb{Z}_X, \mathbb{Z}_X) \simeq \mathbb{Z}.$$

23. Give an example of two sheaves $\mathcal{F}$, $\mathcal{G}$ on a topological space X such that

$$\mathcal{H}om(\mathcal{F}, \mathcal{G})(U) \not\simeq \mathrm{Hom}(\mathcal{F}(U), \mathcal{G}(U))$$

for some open subset $U \subset X$.

24. Let $f : X \to Y$ be a continuous map of topological spaces and let $\mathcal{G}$ be a sheaf of abelian groups on Y. Prove by way of example that

$$U \rightsquigarrow \varinjlim_{U \subset f^{-1}(V)} \mathcal{P}(V)$$

need not be a sheaf.

---

[g]See Exercise 13 in this chapter.

Chapter 4

Cohomology of Sheaves

This chapter is devoted to the description of two cohomology theories associated with a sheaf on a topological space. The first is *Čech cohomology*, which is well suited for studying glueing and extension problems, but on the other hand is not well behaved on a general topological space. This problem is fixed by considering *sheaf cohomology*, which is an example of derived functor.

After introducing these two cohomologies, we shall compare them, and shall discuss applications to the de Rham cohomology of differentiable manifolds, and to the cohomology of quasi-coherent sheaves on schemes.

4.1. Čech Cohomology

In its most natural manifestation, one associates Čech cohomology with a set of data including a topological space X, an open cover of it, and a presheaf $\mathcal{P}$ of abelian groups on X. By definition, a cohomology class of degree p will describe the obstruction to extend a collection of sections of $\mathcal{P}$, defined on intersections of $p+1$ open sets, to sections on intersections of p open sets. Moreover, one can associate Čech cohomology groups just to the pair formed by X and $\mathcal{P}$ by taking a suitable direct limit over the open covers of X. These cohomology groups define functors from the category of sheaves of abelian groups on X to the category of abelian groups; we shall show that when X is paracompact these functors make up a δ-functor. This will allow us to show that when X is paracompact the Čech and sheaf cohomology of a sheaf of abelian groups coincide. However, we shall study in some detail the relation between the two cohomologies also in the general case (i.e., with no restriction on the topological space X). This analysis will be further developed in Chapter 5 by using spectral sequence techniques.

We start by considering a presheaf $\mathcal{P}$ on X and an open cover $\mathfrak{U}$ of X. We assume that $\mathfrak{U}$ is labelled by a totally ordered set I, and define

$$U_{i_0 \cdots i_p} = U_{i_0} \cap \cdots \cap U_{i_p}.$$

We define the *Čech complex of $\mathfrak{U}$ with coefficients in $\mathcal{P}$* as the complex whose pth term is the abelian group

$$C^p(\mathfrak{U}, \mathcal{P}) = \prod_{i_0 < \cdots < i_p} \mathcal{P}(U_{i_0 \cdots i_p}).$$

Thus a p-cochain α is a collection $\{\alpha_{i_0 \cdots i_p}\}$ of sections of $\mathcal{P}$, each one belonging to the space of sections over an intersection of $p + 1$ open sets in $\mathfrak{U}$. Since the indexes of the open sets are taken in strictly increasing order, each intersection is counted only once.

The Čech differential $\delta : C^p(\mathfrak{U}, \mathcal{P}) \to C^{p+1}(\mathfrak{U}, \mathcal{P})$ is defined as follows: if $\alpha = \{\alpha_{i_0 \cdots i_p}\} \in C^p(\mathfrak{U}, \mathcal{P})$, then

$$\{(\delta\alpha)_{i_0 \cdots i_{p+1}}\} = \sum_{k=0}^{p+1} (-1)^k \alpha_{i_0 \cdots \widehat{i_k} \cdots i_{p+1}} |_{U_{i_0 \cdots i_{p+1}}}.$$

Here a caret denotes omission of the index. For instance, if $p = 0$ we have $\alpha = \{\alpha_i\}$ and

$$(\delta\alpha)_{ik} = \alpha_{k|_{U_i \cap U_k}} - \alpha_{i|_{U_i \cap U_k}}. \tag{4.1}$$

It is an easy exercise to check that $\delta^2 = 0$. Thus, we obtain a cohomology theory. We denote the corresponding cohomology groups by $H^k(\mathfrak{U}, \mathcal{P})$.

Lemma 4.1. *If $\mathcal{F}$ is a sheaf, one has an isomorphism $H^0(\mathfrak{U}, \mathcal{F}) \simeq \mathcal{F}(X)$.*

Proof. We have $H^0(\mathfrak{U}, \mathcal{F}) = \ker \delta : C^0(\mathfrak{U}, \mathcal{F}) \to C^1(\mathfrak{U}, \mathcal{F})$. So if $\alpha \in H^0(\mathfrak{U}, \mathcal{F})$ by (4.1) we see that

$$\alpha_{k|_{U_i \cap U_k}} = \alpha_{i|_{U_i \cap U_k}}.$$

By the second sheaf axiom this implies that there is a global section $\tilde{\alpha} \in \mathcal{F}(X)$ such that $\tilde{\alpha}_{|_{U_i}} = \alpha_i$, and by the first sheaf axiom, this is unique. This defines a morphism $H^0(\mathfrak{U}, \mathcal{F}) \to \mathcal{F}(X)$. On the other hand, the morphism $\mathcal{F}(X) \to H^0(\mathfrak{U}, \mathcal{F})$ which maps $\alpha \in \mathcal{F}(X)$ to $\{\alpha_i = \alpha_{|_{U_i}}\}$ inverts the previous one. $\qquad\square$

Example 4.2. We consider an open cover $\mathfrak{U}$ of the circle S^1 formed by three connected sets which intersect only pairwise. We compute the Čech cohomology of $\mathfrak{U}$ with coefficients in the constant sheaf $\mathbb{R}$. We have

$$C^0(\mathfrak{U}, \mathbb{R}) = C^1(\mathfrak{U}, \mathbb{R}) = \mathbb{R} \oplus \mathbb{R} \oplus \mathbb{R},$$

and $C^k(\mathfrak{U}, \mathbb{R}) = 0$ for $k > 1$ because there are no triple intersections. The only nonzero differential $d_0 : C^0(\mathfrak{U}, \mathbb{R}) \to C^1(\mathfrak{U}, \mathbb{R})$ is given by

$$d_0(x_0, x_1, x_2) = (x_1 - x_2, x_2 - x_0, x_0 - x_1).$$

Hence

$$C^0(\mathfrak{U}, \mathbb{R}) = \ker d_0 \simeq \mathbb{R},$$

$$C^1(\mathfrak{U}, \mathbb{R}) = C^1(\mathfrak{U}, \mathbb{R}) / \operatorname{Im} d_0 \simeq \mathbb{R}.$$

It is possible to define Čech cohomology groups depending only on the pair $(X, \mathcal{F})$, and not on a cover, by letting

$$\check{H}^k(X, \mathcal{F}) = \varinjlim_{\mathfrak{U}} H^k(\mathfrak{U}, \mathcal{F}).$$

The direct limit is taken over a cofinal subset of the directed set of all covers of X (the order is of course the refinement of covers: a cover $\mathfrak{V} = \{V_j\}_{j \in J}$ is a refinement of $\mathfrak{U}$ if there is a map $f : J \to I$ such that $V_j \subset U_{f(j)}$ for every $j \in J$). This map must be fixed at the outset, since a cover may be regarded as a refinement of another in many ways. As different cofinal families give rise to the same inductive limit, the groups $\check{H}^k(X, \mathcal{F})$ are well defined.

We wish to show that when X is paracompact,[a] any exact sequence of sheaves induces a corresponding long exact sequence in Čech cohomology. First we show that the long exact cohomology sequence always exists for exact sequences of presheaves, with no assumptions on X; from this, one shows that when X is paracompact, the same is true for exact sequences of sheaves.

Lemma 4.3. *Let X be any topological space, and let*

$$0 \to \mathcal{P}' \to \mathcal{P} \to \mathcal{P}'' \to 0 \qquad (4.2)$$

---

[a]We recall that an open cover $\mathfrak{U}$ of a topological space X is *locally finite* if every point in X has an open neighbourhood which intersects only a finite number of elements of $\mathfrak{U}$. A topological space X is paracompact if it is Hausdorff and any open cover has a locally finite refinement [37].

be an exact sequence of presheaves on X. Then one has a long exact sequence

$$0 \to \check{H}^0(X, \mathcal{P}') \to \check{H}^0(X, \mathcal{P}) \to \check{H}^0(X, \mathcal{P}'') \to \check{H}^1(X, \mathcal{P}') \to$$

$$\cdots \to \check{H}^k(X, \mathcal{P}') \to \check{H}^k(X, \mathcal{P}) \to \check{H}^k(X, \mathcal{P}'') \to \check{H}^{k+1}(X, \mathcal{P}') \to \cdots$$

Proof. For any open cover $\mathfrak{U}$ the exact sequence (4.2) induces an exact sequence of differential complexes

$$0 \to C^\bullet(\mathfrak{U}, \mathcal{P}') \to C^\bullet(\mathfrak{U}, \mathcal{P}) \to C^\bullet(\mathfrak{U}, \mathcal{P}'') \to 0$$

which by Proposition 1.23 induces in turn the long cohomology sequence

$$0 \to H^0(\mathfrak{U}, \mathcal{P}') \to H^0(\mathfrak{U}, \mathcal{P}) \to H^0(\mathfrak{U}, \mathcal{P}'') \to H^1(\mathfrak{U}, \mathcal{P}') \to$$

$$\cdots \to H^k(\mathfrak{U}, \mathcal{P}') \to H^k(\mathfrak{U}, \mathcal{P}) \to H^k(\mathfrak{U}, \mathcal{P}'') \to H^{k+1}(\mathfrak{U}, \mathcal{P}') \to \cdots$$

Since the direct limit of a family of exact sequences yields an exact sequence, by taking the direct limit over the open covers of X one obtains the required exact sequence. $\qquad\square$

Remark 4.4. One can generalize Lemma 4.1 to presheaves, in the following sense. Let $\mathcal{P}$ be a separated presheaf on X (see Definition 3.5), so that the natural morphism $\mathcal{P} \to \mathcal{P}^\natural$ is injective (cf. Exercise 4 in this chapter); instances of this situation were given in Examples 3.20 and 3.21. Then $\check{H}^0(X, \mathcal{P}) \simeq \mathcal{P}^\natural(X)$. We prove this in two steps. First we note that if $\mathcal{Q}$ is a presheaf such that $\mathcal{Q}^\natural = 0$, then $\check{H}^0(X, \mathcal{Q}) = 0$. Indeed, for every open cover $\mathfrak{U}$ of X we have by definition

$$H^0(\mathfrak{U}, \mathcal{Q}) = \ker(\delta_0 : C^0(\mathfrak{U}, \mathcal{Q}) \to C^1(\mathfrak{U}, \mathcal{Q})).$$

Since all stalks of $\mathcal{Q}$ are zero, given a 0-cochain $\alpha = \{\alpha_i\} \in C^0(\mathfrak{U}, \mathcal{Q})$ there is a refinement $\mathfrak{V}$ of $\mathfrak{U}$ such that the image of α in $C^0(\mathfrak{V}, \mathcal{Q})$ is zero (for every point $x \in X$ take a neighbourhood V_{ix} of x such that $(\alpha_i)_{|V_{ix}} = 0$). Taking the direct limit on the open covers, this implies $\check{H}^0(X, \mathcal{Q}) = 0$.

Now let $\mathcal{Q}$ be the quotient presheaf $\mathcal{P}^\natural / \mathcal{P}$, so that the sequence of presheaves

$$0 \to \mathcal{P} \to \mathcal{P}^\natural \to \mathcal{Q} \to 0$$

is exact; then $\mathcal{Q}^\natural = 0$. Taking Čech cohomology of the previous exact sequence we obtain $\check{H}^0(X, \mathcal{P}) \simeq \mathcal{P}^\natural(X)$.

Lemma 4.5. *Let X be a paracompact topological space, $\mathcal{P}$ a presheaf on X whose associated sheaf is the zero sheaf, let $\mathfrak{U}$ be an open cover of X, and* let $\alpha \in C^k(\mathfrak{U}, \mathcal{P})$. *There is a refinement* $\mathfrak{V}$ *of* $\mathfrak{U}$ *such that* $\tau(\alpha) = 0$, *where* $\tau : C^k(\mathfrak{U}, \mathcal{P}) \to C^k(\mathfrak{V}, \mathcal{P})$ *is the morphism induced by restriction.*

Proof. The proof is purely topological. We refer the reader to [31, Lemma 2.9.2] (see also [55, Lemma 2]). $\qquad\square$

Proposition 4.6. *Let* $\mathcal{P}$ *be a presheaf on a paracompact space* X, *and let* $\mathcal{P}^\natural$ *be the associated sheaf. For all* $k \geq 0$, *the natural morphism* $\check{H}^k(X, \mathcal{P}) \to \check{H}^k(X, \mathcal{P}^\natural)$ *is an isomorphism.*

Proof. One has an exact sequence of presheaves

$$0 \to \mathcal{Q}_1 \to \mathcal{P} \to \mathcal{P}^\natural \to \mathcal{Q}_2 \to 0$$

with

$$\mathcal{Q}_1^\natural \simeq \mathcal{Q}_2^\natural = 0. \tag{4.3}$$

This gives rise to

$$0 \to \mathcal{Q}_1 \to \mathcal{P} \to \mathcal{T} \to 0, \quad 0 \to \mathcal{T} \to \mathcal{P}^\natural \to \mathcal{Q}_2 \to 0, \tag{4.4}$$

where $\mathcal{T}$ is the quotient presheaf $\mathcal{P}/\mathcal{Q}_1$, i.e., the presheaf $U \to \mathcal{P}(U)/\mathcal{Q}_1(U)$. By Lemma 4.5, the isomorphisms (4.3) yield $\check{H}^k(X, \mathcal{Q}_1) = \check{H}^k(X, \mathcal{Q}_2) = 0$. Then by taking the long exact sequences of cohomology from the exact sequences (4.4) we obtain the desired isomorphism. $\qquad\square$

Using these results we may eventually prove that on paracompact spaces one has long exact sequences in Čech cohomology.

Theorem 4.7. *Let*

$$0 \to \mathcal{F}' \to \mathcal{F} \to \mathcal{F}'' \to 0$$

be an exact sequence of sheaves on a paracompact space X. *There is a long exact sequence of Čech cohomology groups*

$$0 \to \check{H}^0(X, \mathcal{F}') \to \check{H}^0(X, \mathcal{F}) \to \check{H}^0(X, \mathcal{F}'') \to \check{H}^1(X, \mathcal{F}') \to$$

$$\cdots \to \check{H}^k(X, \mathcal{F}') \to \check{H}^k(X, \mathcal{F}) \to \check{H}^k(X, \mathcal{F}'') \to \check{H}^{k+1}(X, \mathcal{F}') \to \cdots .$$

$$\tag{4.5}$$

Proof. Let $\mathcal{P}$ be the quotient *presheaf* $\mathcal{F}/\mathcal{F}'$; then $\mathcal{P}^\natural \simeq \mathcal{F}''$ (see Example 3.21). One has an exact sequence of presheaves

$$0 \to \mathcal{F}' \to \mathcal{F} \to \mathcal{P} \to 0.$$

By taking the associated long exact sequence in cohomology (cf. Lemma 4.3) and using the isomorphism $\check{H}^k(X, \mathcal{P}) = \check{H}^k(X, \mathcal{F}'')$ from Proposition 4.6, one obtains the required exact sequence. $\qquad\square$

This is the reason why the Čech cohomology groups of X define a δ-functor.

Theorem 4.8. *Let X be a paracompact topological space. The collection of functors*

$$\check{H}^i(X, -) : \mathfrak{Sh}_X \to \mathfrak{Ab}, \quad i \geq 0,$$

together with the connecting morphisms that for each exact sequence of sheaves make the sequence (4.5) exact, is a δ-functor.

Proof. We need to show that for every morphism of exact sequences of sheaves

$$
\begin{array}{ccccccccc}
0 & \longrightarrow & \mathcal{F}' & \longrightarrow & \mathcal{F} & \longrightarrow & \mathcal{F}'' & \longrightarrow & 0 \\
& & \downarrow & & \downarrow & & \downarrow & & \\
0 & \longrightarrow & \mathcal{G}' & \longrightarrow & \mathcal{G} & \longrightarrow & \mathcal{G}'' & \longrightarrow & 0
\end{array}
$$

and for every $i \geq 0$ the square

$$
\begin{array}{ccc}
\check{H}^i(X, \mathcal{F}'') & \longrightarrow & \check{H}^{i+1}(X, \mathcal{F}') \\
\downarrow & & \downarrow \\
\check{H}^i(X, \mathcal{G}'') & \longrightarrow & \check{H}^{i+1}(X, \mathcal{G}')
\end{array}
\qquad (4.6)
$$

commutes.

Let $\mathcal{P} = \mathcal{F}/\mathcal{F}'$ and $\mathcal{Q} = \mathcal{G}/\mathcal{G}'$ be the quotient presheaves. Then one has a morphism of exact sequences of presheaves

$$
\begin{array}{ccccccccc}
0 & \longrightarrow & \mathcal{F}' & \longrightarrow & \mathcal{F} & \longrightarrow & \mathcal{P} & \longrightarrow & 0 \\
& & \downarrow & & \downarrow & & \downarrow & & \\
0 & \longrightarrow & \mathcal{G}' & \longrightarrow & \mathcal{G} & \longrightarrow & \mathcal{Q} & \longrightarrow & 0
\end{array}
$$

Let $\mathfrak{U}$ be an open cover of X; then

$$
\begin{array}{ccccccccc}
0 & \longrightarrow & C^\bullet(\mathfrak{U}, \mathcal{F}') & \longrightarrow & C^\bullet(\mathfrak{U}, \mathcal{F}) & \longrightarrow & C^\bullet(\mathfrak{U}, \mathcal{P}) & \longrightarrow & 0 \\
& & \downarrow & & \downarrow & & \downarrow & & \\
0 & \longrightarrow & C^\bullet(\mathfrak{U}, \mathcal{G}') & \longrightarrow & C^\bullet(\mathfrak{U}, \mathcal{G}) & \longrightarrow & C^\bullet(\mathfrak{U}, \mathcal{Q}) & \longrightarrow & 0
\end{array}
$$

is a morphism of exact sequences of complexes of abelian groups. After Exercise 1.24 this yields for every $i \geq 0$ a commutative square

$$
\begin{array}{ccc}
H^i(\mathfrak{U}, \mathcal{P}) & \longrightarrow & H^{i+1}(\mathfrak{U}, \mathcal{F}') \\
\downarrow & & \downarrow \\
H^i(\mathfrak{U}, \mathcal{Q}) & \longrightarrow & H^{i+1}(\mathfrak{U}, \mathcal{G}')
\end{array}
$$

Taking direct limit over the open covers of X, and, in view of Proposition 4.6, replacing $\mathcal{P}$ and $\mathcal{Q}$ by $\mathcal{F}''$ and $\mathcal{G}''$, respectively, we obtain the square in equation (4.6). $\qquad\qquad\square$

4.2. Sheaf Cohomology

Sheaf cohomology is better behaved than Čech cohomology, as for instance it associates cohomology long exact sequences to exact sequences of sheaves on any topological space. Moreover, sheaf cohomology groups are derived functors (of the global sections functor), so that their study is facilitated by the powerful machinery provided by theory of derived functors.

Ringed spaces and $\mathcal{O}_X$-modules. A convenient category to develop this theory is that of $\mathcal{O}_X$-modules, where $\mathcal{O}_X$ is a sheaf of rings on a topological space X. This will include the category of sheaves of abelian groups as a particular case.

Definition 4.9. A ringed space is a pair $(X, \mathcal{O}_X)$, where X is a topological space, and $\mathcal{O}_X$ is a sheaf of commutative rings with unity on X. A morphism between two locally ringed spaces $(X, \mathcal{O}_X)$ and $(Y, \mathcal{O}_Y)$ is a pair $(f, f^\sharp)$ where $f : X \to Y$ is a continuous map, and $f^\sharp : \mathcal{O}_Y \to f_*\mathcal{O}_X$ is a morphism of sheaves of rings.

Examples of ringed spaces are a topological space with the sheaf of continuous real-valued functions, a differentiable manifold with the sheaf of C^∞ functions, a complex manifold with the sheaf of holomorphic functions, a scheme with the sheaf of regular functions. Also, if X is a topological space and R is a commutative ring with unity, and R_X is the constant sheaf on X with stalk R, then (X, R_X) is a ringed space.

Given a ringed space $(X, \mathcal{O}_X)$, we shall denote by $\mathcal{O}_X$-**mod** the category of sheaves of $\mathcal{O}_X$-modules. Note that a $\mathbb{Z}_X$-module is just a sheaf of abelian groups (i.e., $\mathbb{Z}_X$-**mod** $= \mathfrak{Sh}_X$).[b] We shall denote by $\mathrm{Hom}_{\mathcal{O}_X}$ the Hom groups in the category $\mathcal{O}_X$-**mod**.

We need to extend the direct and inverse images functors we defined in Section 3.4 to the category of $\mathcal{O}_X$-modules. First, one easily checks that, if $f : (X, \mathcal{O}_X) \to (Y, \mathcal{O}_Y)$ is a morphism of ringed spaces, and $\mathcal{F}$ is an $\mathcal{O}_X$-module, the direct image $f_*\mathcal{F}$, as defined in Section 3.4, has a natural structure of $\mathcal{O}_Y$-module. On the other hand, if $\mathcal{G}$ is an $\mathcal{O}_Y$-module, the inverse image $f^{-1}\mathcal{G}$ is not an $\mathcal{O}_X$-module. However, $\mathcal{O}_X$ does have an $\mathcal{O}_Y$-module structure, or more precisely, an $f^{-1}\mathcal{O}_Y$-module structure, obtained in the following way: as part of the data provided by the morphism f, one has a morphism $f^\sharp : \mathcal{O}_Y \to f_*\mathcal{O}_X$, which by the adjunction (3.2) produces a morphism $f^{-1}\mathcal{O}_Y \to \mathcal{O}_X$. This makes $\mathcal{O}_X$ into an $f^{-1}\mathcal{O}_Y$-module, and we can use this to make $f^{-1}\mathcal{G}$ into an $\mathcal{O}_X$-module by defining

$$f^*\mathcal{G} = f^{-1}\mathcal{G} \otimes_{f^{-1}\mathcal{O}_Y} \mathcal{O}_X.$$

Then the adjunction between the inverse and direct images of abelian groups expressed in equation (3.2) extends to an adjunction

$$\mathrm{Hom}_{\mathcal{O}_X}(f^*\mathcal{G}, \mathcal{F}) \simeq \mathrm{Hom}_{\mathcal{O}_Y}(\mathcal{G}, f_*\mathcal{F}). \tag{4.7}$$

This may be proved from (3.2) by applying the Tensor-Hom adjunction (see Exercise 13 in Chapter 2).

Sheaf cohomology. Since we shall define sheaf cohomology groups as the derived functors of the global section functor, we need to check that the category of $\mathcal{O}_X$-modules has enough injectives.

Proposition 4.10. *The category $\mathcal{O}_X$-**mod** has enough injectives.*

Proof. Let $\mathcal{F}$ be an $\mathcal{O}_X$-module; for every $x \in X$ there is an immersion $i_x : \mathcal{F}_x \to \mathcal{I}_x$ of the stalk $\mathcal{F}_x$ into an injective $\mathcal{O}_{X,x}$-module $\mathcal{I}_x$. Let $j_x : \{x\} \to X$ be the inclusion of the one-point space $\{x\}$ into X, and define

$$\mathcal{I} = \prod_{x \in X} j_{x,*}\mathcal{I}_x.$$

We consider $(\{x\}, \mathcal{O}_{X,x})$ as a ringed space; then the adjunction (4.7) yields

$$\mathrm{Hom}_{\mathcal{O}_X}(\mathcal{F}, j_{x,*}\mathcal{I}_x) \simeq \mathrm{Hom}_{\mathcal{O}_{X,x}}(\mathcal{F}_x, \mathcal{I}_x),$$

so that $i = \prod_{x \in X} i_x$ is an injection $i : \mathcal{F} \to \mathcal{I}$. Now we need to show that $\mathcal{I}$ is an injective object in $\mathcal{O}_X$-**mod**. Indeed for every $\mathcal{O}_X$-module

---

[b]This identification is not entirely trivial, for a discussion see https://mathoverflow.net/questions/67709/sheaves-of-mathbb-z-modules-sheaves-of-abelian-groups.

$\mathcal{F}$ we have

$$\text{Hom}_{\mathcal{O}_X}(\mathcal{F},\mathcal{I}) = \prod_{x \in X} \text{Hom}_{\mathcal{O}_{X,x}}(S_x(\mathcal{F}),\mathcal{I}_x),$$

where $S_x : \mathcal{O}_X\text{-mod} \to \mathcal{O}_{X,x}\text{-mod}$ is the functor which takes the stalk of an $\mathcal{O}_X$-module at x. Both $\text{Hom}_{\mathcal{O}_{X,x}}(-,\mathcal{I}_x)$ and S_x are exact, in particular because $\mathcal{I}_x$ is injective, so that $\text{Hom}_{\mathcal{O}_X}(-,\mathcal{I})$ is exact, which means that $\mathcal{I}$ is injective. $\qquad\square$

We regard the operation of taking the global sections of an $\mathcal{O}_X$-module as a functor

$$\Gamma : \mathcal{O}_X\text{-mod} \to \Gamma(X,\mathcal{O}_X)\text{-mod}. \qquad (4.8)$$

This is clearly additive, and moreover it is left-exact (cf. Remark 3.7), so that we can take its right derived functors.

Definition 4.11 (Sheaf cohomology). The sheaf cohomology groups are the right derived functors of the global section functor (4.8); if $\mathcal{F}$ is an $\mathcal{O}_X$-module, we denote

$$H^i(X,\mathcal{F}) = R^i\Gamma(\mathcal{F}), \quad i \geq 0.$$

Note that $H^0(X,\mathcal{F}) = \Gamma(X,\mathcal{F})$ (cf. Remark 2.13).

In view of Theorem 2.17, given an exact sequence of $\mathcal{O}_X$-modules there is a long exact sequence of cohomology.

Proposition 4.12. *If* $0 \to \mathcal{F}' \to \mathcal{F} \to \mathcal{F}'' \to 0$ *is an exact sequence of* $\mathcal{O}_X$*-modules, there is a long exact sequence*

$$0 \to H^0(X,\mathcal{F}') \to H^0(X,\mathcal{F}) \to H^0(X,\mathcal{F}'') \to H^1(X,\mathcal{F}')$$
$$\to H^1(X,\mathcal{F}) \to H^1(X,\mathcal{F}'') \to H^2(X,\mathcal{F}') \to \cdots.$$

Since, as we have noted above, an $\mathcal{O}_X$-module is also a sheaf of abelian groups, i.e., a $\mathbb{Z}_X$-module, one could take its sheaf cohomology groups as such, deriving the functor

$$\Gamma : \mathfrak{Sh}_X \to \mathfrak{Ab}.$$

More precisely, there are forgetful functors

$$\mathcal{O}_X\text{-mod} \to \mathfrak{Sh}_X \quad \text{and} \quad \Gamma(X,\mathcal{O}_X)\text{-mod} \to \mathfrak{Ab}$$

fitting into a commutative diagram of functors

$$
\begin{array}{ccc}
\mathcal{O}_X\text{-}\mathbf{mod} & \xrightarrow{\ \Gamma\ } & \Gamma(X,\mathcal{O}_X)\text{-}\mathbf{mod} \\
\downarrow & & \downarrow \\
\mathfrak{Sh}_X & \xrightarrow{\ \ \Gamma\ \ } & \mathfrak{Ab}
\end{array}
$$

and we are asking what is the action of these forgetful functors on sheaf cohomology. It turns out that, given an $\mathcal{O}_X$-module, its sheaf cohomology groups as an $\mathcal{O}_X$-module or as a sheaf of abelian groups are canonically isomorphic. To show this, we introduce the notion of *flabby* sheaf,[c] which will allow us to realize sheaf cohomology by means of Γ-acyclic resolutions.

Definition 4.13 (Flabby sheaves). A sheaf of $\mathcal{O}_X$-modules $\mathcal{F}$ is flabby if for every inclusion of open sets $V \subset U$, the restriction morphism $\mathcal{F}(U) \to \mathcal{F}(V)$ is surjective.

Example 4.14. Given a topological space X and an abelian group G, and after fixing a point $x \in X$, one defines the *skyscraper sheaf* $G(x)$ as

$$
G(x)(U) = \begin{cases} G & \text{if } x \in U, \\ 0 & \text{if } x \notin U. \end{cases}
$$

Quite clearly, the sheaf $G(x)$ is flabby. Its stalk is 0 everywhere but at the points of the closure of x, where it is G.

Example 4.15. If $\mathcal{F}$ is a sheaf, the sheaf of all sections of the étalé space $\underline{\mathcal{F}}$, continuous or not, is flabby — just extend sections by zero (see Remark 4.23).

Exercise 4.16. As we mentioned in Exercise 13 in Chapter 3, a nonempty topological space X is *irreducible* if it cannot be written as the union of two proper closed subsets. Prove that a constant sheaf on an irreducible topological space is flabby.

---

[c] In [28] flabby sheaves are called "flasque", the French term for them.

Example 4.17 (An irreducible topological space). Let X be a set with three elements, $X = \{x, y, z\}$, and declare its open subsets to be

$$U_0 = \emptyset, \quad U_1 = \{x\}, \quad U_2 = \{x, y\}, \quad U_3 = X.$$

Then X is irreducible.

We develop some preliminary results.

Proposition 4.18. *If* $0 \to \mathcal{F}' \xrightarrow{i} \mathcal{F} \xrightarrow{p} \mathcal{F}'' \to 0$ *is an exact sequence of sheaves and* $\mathcal{F}'$ *is flabby, then the sequence is exact also as a sequence of presheaves.*

Proof. We need to prove that $p_U : \mathcal{F}(U) \to \mathcal{F}''(U)$ is surjective for all open sets $U \subset X$. Let $s'' \in \mathcal{F}''(U)$ and let S be the set of pairs (W, s), where $W \subset U$ is open and $s \in \mathcal{F}(W)$ is such that $p_W(s) = s''_{|W}$. The set S is nonempty since the sequence is exact as sheaves. It is partially ordered by

$$(W, t) < (W', t') \quad \text{if} \quad W \subset W' \quad \text{and} \quad t = t'_{|W}.$$

Any nonempty chain in the set S has an upper bound (one takes the union of all open sets W, and checks that all sections in the pairs (W, s) extend to a section over the union), hence by Zorn's Lemma[d] S has a maximal element, say $(\overline{W}, \overline{t})$.

We prove that $U = \overline{W}$. This shows the surjectivity of p_U. Assume $x \in U - \overline{W}$. Then there is an open neighbourhood V of x and element $t \in \mathcal{F}(V)$ such that $p_V(t) = s''_{|V}$. Note that $\overline{t}_{|\overline{W} \cap V} - t_{|\overline{W} \cap V} \in \mathcal{F}'(\overline{W} \cap V)$ so since $\mathcal{F}'$ is flabby, the difference can be extended to V: there is $u \in \mathcal{F}'(V)$ such that $i_V(u)_{|\overline{W} \cap V} = \overline{t}_{|\overline{W} \cap V} - t_{|\overline{W} \cap V}$. Let $t' = t + i_V(u) \in \mathcal{F}(V)$.

Let $\widetilde{W} = \overline{W} \cup V$; then $t' \in \mathcal{F}(V)$ and $\overline{t} \in \mathcal{F}(\overline{W})$ glue to a section $\tilde{s}$ in $\mathcal{F}(\widetilde{W})$. Indeed

$$t'_{|\overline{W} \cap V} = \overline{t}_{|\overline{W} \cap V}.$$

So $(\overline{W}, \overline{t}) < (\widetilde{W}, \tilde{s})$ properly, which contradicts the maximality of $(\overline{W}, \overline{t})$.
$\square$

---

[d] Zorn's Lemma stipulates that a nonempty partially ordered set where every nonempty chain (a totally ordered subset) has an upper bound, always has at least one maximal element [27].

Corollary 4.19. *A quotient of flabby sheaves is flabby.*

Proof. Let $V \subset U$ be an inclusion of open subsets of X, and consider the commutative diagram

$$
\begin{array}{ccccccccc}
0 & \longrightarrow & \mathcal{F}'(U) & \longrightarrow & \mathcal{F}(U) & \longrightarrow & \mathcal{F}''(U) & \longrightarrow & 0 \\
 & & \downarrow & & \downarrow & & \downarrow & & \\
0 & \longrightarrow & \mathcal{F}'(V) & \longrightarrow & \mathcal{F}(V) & \longrightarrow & \mathcal{F}''(V) & \longrightarrow & 0 \\
 & & \downarrow & & \downarrow & & \downarrow & & \\
 & & 0 & & 0 & & 0 & &
\end{array}
$$

Its rows are exact by Proposition 4.18, and the middle vertical arrow is surjective as $\mathcal{F}$ is flabby; then also the third vertical arrow is surjective. $\qquad \square$

Proposition 4.20. *Every injective $\mathcal{O}_X$-module is flabby.*

Proof. Let $\mathcal{I}$ be an injective $\mathcal{O}_X$-module. For every open set $U \subset X$ and $\mathcal{O}_U$-module $\mathcal{F}$ define an $\mathcal{O}_X$-module $j_!(\mathcal{F})$ by letting

$$
j_!(\mathcal{F})(V) = \begin{cases} \mathcal{F}(V) & \text{if } V \subset U, \\ 0 & \text{if } V \not\subset U. \end{cases} \tag{4.9}
$$

We shall denote $\mathcal{O}_{(U)} = j_!(\mathcal{O}_U)$. If $V \subset U$ are nested open subsets of X, we have an injective morphism $\mathcal{O}_{(V)} \to \mathcal{O}_{(U)}$; by applying $\mathrm{Hom}_{\mathcal{O}_X}(-, \mathcal{I})$ to the inclusion $\mathcal{O}_{(V)} \to \mathcal{O}_{(U)}$ we get a surjection $\mathrm{Hom}_{\mathcal{O}_X}(\mathcal{O}_{(U)}, \mathcal{I}) \to \mathrm{Hom}_{\mathcal{O}_X}(\mathcal{O}_{(V)}, \mathcal{I})$. Since $\mathrm{Hom}_{\mathcal{O}_X}(\mathcal{O}_{(U)}, \mathcal{I}) \simeq \mathcal{I}(U)$, and the same for V, we get a surjection $\mathcal{I}(U) \to \mathcal{I}(V)$, i.e., $\mathcal{I}$ is flabby. $\qquad \square$

Remark 4.21. By taking $\mathcal{O}_X = \mathbb{Z}_X$, one sees that injective sheaves of abelian groups are flabby.

Theorem 4.22. *Flabby sheaves of $\mathcal{O}_X$-modules are acyclic for sheaf cohomology. So they compute sheaf cohomology.*

Proof. Let $\mathcal{F}$ be a flabby $\mathcal{O}_X$-module and embed it into an injective $\mathcal{O}_X$-module $\mathcal{I}$. By Corollary 4.19 and Proposition 4.20, the quotient $\mathcal{Q} = \mathcal{I}/\mathcal{F}$ is flabby. Taking cohomology, and since $\mathcal{I}$ is acyclic, we have (cf. Proposition 4.12)

$$
0 \to H^0(X, \mathcal{F}) \to H^0(X, \mathcal{I}) \to H^0(X, \mathcal{Q}) \to H^1(X, \mathcal{F}) \to 0
$$

$$
0 \to H^i(X, \mathcal{Q}) \xrightarrow{\sim} H^{i+1}(X, \mathcal{F}) \to 0, \quad i \geq 1.
$$

By Proposition 4.18, as $\mathcal{F}$ is flabby, the sequence of the H^0 is exact, so that $H^1(X, \mathcal{F}) = 0$. Since $\mathcal{Q}$ is flabby, the second sequence allows us to make induction, proving that $H^i(X, \mathcal{F}) = 0$ for $i > 0$.   □

Remark 4.23. Sheaves of abelian groups have a canonical flabby resolution, called *Godement resolution* [20]. Let $\mathcal{F}$ be a sheaf of abelian groups, and denote by $\mathcal{C}^0(\mathcal{F})$ the sheaf of all sections, continuous or not, of the étalé space $\underline{\mathcal{F}}$ (see Section 3.2). As we noted in Example 4.15, the sheaf $\mathcal{C}^0(\mathcal{F})$ is flabby, and there is an injection $\mathcal{F} \to \mathcal{C}^0(\mathcal{F})$, as $\mathcal{F}$ is the sheaf of continuous sections of $\underline{\mathcal{F}}$. Proceeding as in Proposition 2.10 we can construct a resolution $\mathcal{C}^\bullet(\mathcal{F})$ by flabby sheaves.

Remark 4.24. If $\mathcal{F}$ is an $\mathcal{O}_X$-module, we may look at it as a sheaf of abelian groups, and resolve it by a flabby resolution (for instance, the Godement resolution); by Theorem 4.22, the resulting cohomology groups are isomorphic to those we get by using injective resolutions in the category $\mathcal{O}_X$-mod.

Higher direct images. In Section 3.4, given a continuous map of topological spaces $f : X \to Y$, we defined the left exact functor $f_* : \mathfrak{Sh}_X \to \mathfrak{Sh}_Y$, called the *direct image functor*. Since the category $\mathfrak{Sh}_X$ has enough injectives (as a particular case of Proposition 4.10), we can introduce the *higher direct images*

$$R^i f_* : \mathfrak{Sh}_X \to \mathfrak{Sh}_Y, \quad i \geq 0$$

as its right derived functors. As usual we have $R^0 f_* \simeq f_*$.

The sheaves $R^i f_* \mathcal{F}$ may be regarded as expressing the "relative cohomology" of $\mathcal{F}$, as the following more explicit characterization will show. Let us define the functors

$$S_f^i : \mathfrak{Sh}_X \to \mathfrak{Sh}_Y, \quad i \geq 0,$$

where $S_f^i(\mathcal{F})$ is the sheaf associated to the presheaf over Y

$$U \rightsquigarrow H^i(f^{-1}(U), \mathcal{F}).$$

Actually, for every $i \geq 0$ the functors $R^i f_*$ and S_f^i are isomorphic. This holds true because both collections of functors are universal δ-functors,

and they coincide in degree zero. The steps to prove this isomorphism are
the following:

- $S^0_f \simeq f_* \simeq R^0 f_*$ basically by definition.
- The properties of the cohomology long exact sequence associated with an
  exact sequence of sheaves imply that the functors S^i_f make up a δ-functor
  (cf. Section 2.6).
- This δ-functor is effaceable (cf. Definition 2.25 and Lemma 2.26). Indeed,
  if $\mathcal{I}$ is an injective sheaf, by Proposition 4.20 it is flabby, and then $S^i_f(\mathcal{I}) =$
  0 when $i > 0$ by Theorem 4.22.
- Finally, $S^i_f \simeq R^i f_*$ by Corollary 2.30.

The interpretation of the functors $R^i f_*$ as relative cohomology functors
is further supported by the following result, whose proof can be found
in [20, Remark. 4.17.1]. For $y \in Y$ denote by $F(y)$ the fibre $f^{-1}(y) \subset X$
equipped with the relative topology. Since a presheaf and the associated
sheaf have the same stalks, we have

$$(R^i f_* \mathcal{F})_y \simeq \varinjlim_{x \in U} H^i(f^{-1}(U), \mathcal{F})$$

(here $(R^i f_* \mathcal{F})_y$ is the stalk of $R^i f_* \mathcal{F}$ at y), so that for every $i \geq 0$ there is
a morphism

$$\phi^i : (R^i f_* \mathcal{F})_y \to H^i(F(y), \mathcal{F})$$

(represent a germ in $(R^i f_* \mathcal{F})_y$ with an element in $H^i(f^{-1}(U), \mathcal{F})$ for some
open neighbourhood U of y and restrict the element to $F(y)$ using the
restriction morphisms defined in Exercise 5 in this chapter).

Proposition 4.25. *If X and Y are locally compact,[e] and f is proper, the
morphisms ϕ^i are isomorphisms for all $i \geq 0$.*

4.3. Comparing Čech and Sheaf Cohomology

We have so far introduced two cohomologies for sheaves, namely, Čech
cohomology and sheaf cohomology. It is natural to ask how these cohomologies are related. There are actually two geometric situations where the
two cohomologies coincide, namely, when the base space X is paracompact
(for sheaves of abelian groups), or when it is a scheme satisfying suitable

---

[e]A topological space is locally compact if every point has a compact neighbourhood.

conditions (for sheaves of $\mathcal{O}_X$-modules). We begin by considering the first situation.

For every topological space X and sheaf $\mathcal{F}$ of abelian groups on it, and every open cover $\mathfrak{U}$ of X, there are morphisms

$$H^i(\mathfrak{U}, \mathcal{F}) \to H^i(X, \mathcal{F}), \quad i \geq 0 \tag{4.10}$$

between the Čech cohomology groups of $\mathfrak{U}$ with coefficients in $\mathcal{F}$, and the sheaf cohomology of $\mathcal{F}$. By taking direct limit over the open covers of X, one obtains morphisms $\check{H}^i(X, \mathcal{F}) \to H^i(X, \mathcal{F})$. We shall construct these morphisms by using the *Čech resolution of $\mathcal{F}$*.

We shall use the notation of Section 4.1. Let $j_{i_0 \cdots i_p} : U_{i_0 \cdots i_p} \to X$ be the inclusion of a nonvoid intersection $U_{i_0 \cdots i_p}$ of elements in the open cover $\mathfrak{U}$. For every p define the sheaf

$$\check{C}^p(\mathfrak{U}, \mathcal{F}) = \prod_{i_0 < \cdots < i_p} (j_{i_0 \cdots i_p})_* \mathcal{F}|_{U_{i_0 \cdots i_p}}$$

(every factor $(j_{i_0 \cdots i_p})_* \mathcal{F}|_{U_{i_0 \cdots i_p}}$ is the direct image on the whole of X of the restriction of $\mathcal{F}$ to $U_{i_0 \cdots i_p}$). More explicitly, the group of sections of the sheaf $\check{C}^p(\mathfrak{U}, \mathcal{F})$ over an open set $U \subset X$ is

$$\check{C}^p(\mathfrak{U}, \mathcal{F})(U) = \prod_{i_0 < \cdots < i_p} \mathcal{F}(U \cap U_{i_0} \cap \cdots \cap U_{i_p}).$$

The usual Čech differential induces sheaf morphisms

$$\delta : \check{C}^p(\mathfrak{U}, \mathcal{F}) \to \check{C}^{p+1}(\mathfrak{U}, \mathcal{F}),$$

which square to zero.

Note that there is a morphism $\epsilon : \mathcal{F} \to \check{C}^0(\mathfrak{U}, \mathcal{F})$ that for every open set $U \subset X$ maps $s \in \mathcal{F}(U)$ to $\prod_i s|_{U \cap U_i} \in \check{C}^0(\mathfrak{U}, \mathcal{F})(U)$.

The next proposition shows some properties of the sheaf complex $(\check{C}^\bullet(\mathfrak{U}, \mathcal{F}), \delta)$.

Proposition 4.26.

(1) *There are isomorphisms*

$$\Gamma(X, \check{C}^p(\mathfrak{U}, \mathcal{F})) \simeq C^p(\mathfrak{U}, \mathcal{F}),$$

i.e., by taking global sections of the Čech sheaf complex we get the Čech cochain group complex.

(2) *For all p and k,*

$$\check{H}^k(X, \check{C}^p(\mathfrak{U}, \mathcal{F})) \simeq \prod_{i_0 < \cdots < i_p} \check{H}^k(U_{i_0 \cdots i_p}, \mathcal{F}).$$

(3) *The complex $\check{C}^\bullet(\mathfrak{U}, \mathcal{F})$ is a resolution of $\mathcal{F}$, i.e., the sequence of sheaves*

$$0 \to \mathcal{F} \xrightarrow{\epsilon} \check{C}^0(\mathfrak{U}, \mathcal{F}) \xrightarrow{\delta_0} \check{C}^1(\mathfrak{U}, \mathcal{F}) \xrightarrow{\delta_1} \check{C}^2(\mathfrak{U}, \mathcal{F}) \to \cdots$$

is exact.

Proof. The first claim is evident. To prove the second, we remind that by the definition of the Čech cohomology groups we have

$$\check{H}^k(X, \check{C}^p(\mathfrak{U}, \mathcal{F})) = \varinjlim_{\mathfrak{V}} H^k(\mathfrak{V}, \check{C}^p(\mathfrak{U}, \mathcal{F})),$$

where $\mathfrak{V}$ runs over all open covers of X. The groups $H^k(\mathfrak{V}, \check{C}^p(\mathfrak{U}, \mathcal{F}))$ are the cohomology of the complex $C^\bullet(\mathfrak{V}, \check{C}^p(\mathfrak{U}, \mathcal{F}))$, which may be written as

$$C^k(\mathfrak{V}, \check{C}^p(\mathfrak{U}, \mathcal{F})) = \prod_{\ell_0 < \cdots < \ell_k} \check{C}^p(\mathfrak{U}, \mathcal{F})(V_{\ell_0 \cdots \ell_k})$$

$$\simeq \prod_{\ell_0 < \cdots < \ell_k} \prod_{i_0 < \cdots < i_p} \mathcal{F}(V_{\ell_0 \cdots \ell_k} \cap U_{i_0 \cdots i_p})$$

$$\simeq \prod_{i_0 < \cdots < i_p} C^k(\mathfrak{V}, (j_{i_0 \cdots i_p})_* \mathcal{F}|_{U_{i_0 \cdots i_p}}).$$

Taking cohomology and direct limit on the cover $\mathfrak{V}$ one obtains the claim. Concerning the third claim, the isomorphism

$$\mathcal{F} \simeq \ker(\delta_0 : \check{C}^0(\mathfrak{U}, \mathcal{F}) \to \check{C}^1(\mathfrak{U}, \mathcal{F}))$$

follows from the sheaf axioms, as in Lemma 4.1. To prove exactness in positive degree, we construct, for every $x \in X$, a null homotopy for the complex of stalks $\check{C}^\bullet(\mathfrak{U}, \mathcal{F})_x$ (cf. Proposition 1.30), i.e., for $p \geq 1$, a morphism

$$K_x : \check{C}^p(\mathfrak{U}, \mathcal{F})_x \to \check{C}^{p-1}(\mathfrak{U}, \mathcal{F})_x.$$

Suppose $x \in U_j$. A germ $s_x \in \check{C}^p(\mathfrak{U}, \mathcal{F})_x$ is represented by a section $s \in \check{C}^p(\mathfrak{U}, \mathcal{F})(V)$, and we can assume $V \subset U_j$. Now let

$$(Ks)_{i_0 \cdots i_{p-1}} = (-1)^\sigma s_{i_0 \cdots j \cdots i_{p-1}},$$

where σ is the permutation that sends $\{j, i_0, \ldots, i_{p-1}\}$ to the sequence of ordered indexes $\{i_0, \ldots, j, \ldots, i_{p-1}\}$; finally, let

$$K_x(s_x) = \prod_{i_0 < \cdots < i_{p-1}} [(Ks)_{i_0 \cdots i_{p-1}}]_x.$$

A direct calculation shows that

$$\delta \circ K_x + K_x \circ \delta = \text{id}. \qquad \square$$

From the fact that $\check{C}^\bullet(\mathfrak{U}, \mathcal{F})$ is a resolution of $\mathcal{F}$, and from Proposition 2.11, we know there exists a morphism of complexes $\check{C}^\bullet(\mathfrak{U}, \mathcal{F}) \to \mathcal{I}^\bullet$, where $\mathcal{I}^\bullet$ is an injective resolution of $\mathcal{F}$, which lifts the identity morphism $\mathcal{F} \to \mathcal{F}$. If we apply the global section functor Γ to this morphism of complexes, in view of item (1) of Proposition 4.26 we obtain morphisms as in (4.10). Note that for $i = 0$ the morphism is an isomorphism as Γ is left exact, and both $H^0(\mathfrak{U}, \mathcal{F})$ and $H^0(X, \mathcal{F})$ are identified with $\Gamma(X, \mathcal{F})$. In general, taking direct limit over the open cover, one obtains morphisms

$$\check{H}^i(X, \mathcal{F}) \to H^i(X, \mathcal{F}), \quad i \geq 0; \tag{4.11}$$

again, in degree 0 this is an isomorphism. However, we cannot always expect these morphisms to be isomorphisms, as the following example shows.

We shall need an easy lemma.

Lemma 4.27. *If $i : Z \to X$ is the inclusion of a closed subset, and $\mathcal{G}$ a sheaf of abelian groups on Z, then $H^k(Z, \mathcal{G}) \simeq H^k(X, i_*\mathcal{G})$ for $k \geq 0$.*

Proof. The proof is based on the following two facts, whose verification we leave to the reader (see also Exercise 18 in Chapter 3):

(1) the functor $i_* : \mathfrak{Sh}_Z \to \mathfrak{Sh}_X$ is exact;
(2) if $\mathcal{L}$ is a flabby sheaf on Z, then $i_*\mathcal{L}$ is flabby.

This implies that if $\mathcal{L}^\bullet$ is a flabby resolution of a sheaf $\mathcal{G}$ on Z, then $i_*\mathcal{L}^\bullet$ is a flabby resolution of $i_*\mathcal{G}$. Then we have

$$H^k(X, i_*\mathcal{G}) \simeq H^k(\Gamma(X, i_*\mathcal{L}^\bullet)) = H^k(\Gamma(Z, \mathcal{L}^\bullet)) \simeq H^k(Z, \mathcal{G}). \qquad \square$$

Example 4.28. We will see here that the second Čech cohomology group is not isomorphic, in general, to the second sheaf cohomology group. This is taken from [22, Section 3.8].

Let X be an irreducible space; for simplicity, we may assume that X is the affine plane over a field $\Bbbk$, Y_1 and Y_2 are two irreducible closed subspaces of X such that $Y_1 \cap Y_2 = \{p, q\}$. We shall denote by $\Bbbk$ the constant sheaf with stalk a field $\Bbbk$ on X. If $Y = Y_1 \cup Y_2$, one has the following exact sequence (cf. Exercise 18 in Chapter 3):

$$0 \to j_!(\Bbbk_{|X-Y}) \to \Bbbk \to i_*\Bbbk_{|Y} \to 0, \qquad (4.12)$$

where $j : X - Y \to X$ is the inclusion of the open subset $X - Y$ in X, the first sheaf in the exact sequence is the one defined in (4.9), and $i : Y \to X$ is the inclusion of the closed subset Y in X. We will prove in Example 5.35 that $\check{H}^2(X, j_!(\Bbbk_{|X-Y})) = 0$, while we are going to see here that $H^2(X, j_!(\Bbbk_{|X-Y})) = \Bbbk$. By Exercise 4.16, the constant sheaf $\Bbbk$ is flabby, so that $H^i(X, \Bbbk) = 0$ for all $i > 0$. Therefore, from the long exact sequence of cohomology we have $H^2(X, j_!(\Bbbk_{|X-Y})) = H^1(X, i_*\Bbbk_{|Y})$ and by Lemma 4.27, $H^1(X, i_*\Bbbk_{|Y}) = H^1(Y, \Bbbk_Y)$, where $\Bbbk_Y$ is the constant sheaf on Y.

We claim that $H^1(Y, \Bbbk_Y) = \Bbbk$. Indeed, one has an exact sequence (cf. Exercise 20 in Chapter 3)

$$0 \to \Bbbk_Y \to i_{1,*}\Bbbk_{|Y_1} \oplus i_{2,*}\Bbbk_{|Y_2} \to i_{12,*}\Bbbk_{|Y_1 \cap Y_2} \to 0,$$

where the direct images are given by the inclusions of the corresponding closed subsets of Y. Again, by Lemma 4.27, $H^k(Y, i_{\alpha,*}\Bbbk_{|Y_1})$, with $\alpha = 1, 2$, coincides with the cohomology group of the constant sheaf over the irreducible closed subset Y_1, and thus it is zero for all $k > 0$, and the same for Y_2. Therefore,

$$0 \to \Gamma(Y, \Bbbk_Y) \to \Gamma(Y, i_{1,*}\Bbbk_{|Y_1} \oplus i_{2,*}\Bbbk_{|Y_2})$$
$$\to \Gamma(Y, i_{12,*}\Bbbk_{|Y_1 \cap Y_2}) \to H^1(Y, \Bbbk_Y) \to 0$$

and $H^1(Y, \Bbbk_Y) = \Bbbk$, as Y, Y_1 and Y_2 are connected.

Nevertheless, on paracompact spaces the two cohomologies do coincide. Indeed, as we noted in Section 4.1, when X is paracompact, the functors $\check{H}^i(X, -) : \mathfrak{Sh}_X \to \mathfrak{Ab}$ make up a δ-functor. This δ-functor happens to be universal, as a consequence of the following lemma (note however that this result holds true on any topological space).

Lemma 4.29. *Let $\mathcal{F}$ be a flabby sheaf of abelian groups on a topological space X. For any open cover $\mathfrak{U}$ of X, one has $H^i(\mathfrak{U}, \mathcal{F}) = 0$ for $i > 0$. Therefore, taking direct limit over the open covers, we also have $\check{H}^i(X, \mathcal{F}) = 0$ for $i > 0$.*

Proof. It is quite clear that the sheaves $\check{C}^i(\mathfrak{U}, \mathcal{F})$ are flabby as well, and therefore Γ-acyclic (i.e., they have vanishing sheaf cohomology in positive degree, see Theorem 4.22). As a result, the resolution $\check{C}^\bullet(\mathfrak{U}, \mathcal{F})$ computes the sheaf cohomology of $\mathcal{F}$, which is zero in positive degree. On the other hand, by Proposition 4.26,

$$H^i(\Gamma(X, \check{C}^\bullet(\mathfrak{U}, \mathcal{F}))) \simeq H^i(C^\bullet(\mathfrak{U}, \mathcal{F})) = H^i(\mathfrak{U}, \mathcal{F}). \qquad \square$$

Example 4.30. Let $\Bbbk_X$ be the constant sheaf on the irreducible topological space X of Example 4.17 with stalk a field $\Bbbk$. As $\Bbbk_X$ is flabby (Exercise 4.16), by Lemma 4.29 its Čech cohomology is trivial for any open cover of X. The reader can check this by direct computation for the open cover $\mathfrak{U} = \{U_1, U_2, U_3\}$.

Theorem 4.31. *If X is paracompact, then $\check{H}^i(X, \mathcal{F}) \simeq H^i(X, \mathcal{F})$ for all sheaves of abelian groups $\mathcal{F}$ on X.*

Proof. Let $\mathcal{I}$ be an injective sheaf of abelian groups on X; by Proposition 4.20, $\mathcal{I}$ is flabby, hence by Lemma 4.29 it has vanishing Čech cohomology in positive degree.[f] By Lemma 2.26 and Theorem 2.29, the δ-functor $\{\check{H}^i(X, -)\}$ is universal. Moreover, by Lemma 4.1 we have

$$\check{H}^0(X, \mathcal{F}) \simeq \Gamma(X, \mathcal{F}) \simeq H^0(X, \mathcal{F})$$

as sheaf cohomology groups are the derived functors of the global sections functor. Then we conclude by Corollary 2.30. $\qquad \square$

Remark 4.32. We know that the morphism (4.11) is an isomorphism for $i = 0$. Actually, for any topological space X, it is an isomorphism also for $i = 1$, and it is injective for $i = 2$. Let us show these two facts.[g] Given a sheaf of abelian groups $\mathcal{F}$, let us embed it into a flabby sheaf $\mathcal{G}$ (cf. Remark 4.23), and let $\mathcal{Q}$ be the quotient presheaf $\mathcal{G}/\mathcal{F}$; then the associated sheaf $\mathcal{Q}^\natural$ is the quotient sheaf. Since $\mathcal{G}$ is flabby, we have $H^1(X, \mathcal{G}) = H^1(\mathfrak{U}, \mathcal{G}) = 0$ (Theorem 4.22 and Lemma 4.29). By taking the long exact sequence of

---

[f]One can also directly prove that injective sheaves have vanishing Čech cohomology by defining a null homotopy for the Čech complex, see, e.g., [59, Lemma 4.12].
[g]This is the proof suggested in [28, Exercise III.4.4]. There is also a proof which uses spectral sequences, see Section 5.6.3.

cohomology associated with the exact sequence of presheaves

$$0 \to \mathcal{F} \to \mathcal{G} \to \mathcal{Q} \to 0$$

with respect to Čech cohomology, and the long exact sequence for sheaf cohomology associated with the exact sequence of sheaves

$$0 \to \mathcal{F} \to \mathcal{G} \to \mathcal{Q}^\natural \to 0,$$

we can form the commutative diagram[h]

$$
\begin{array}{ccccccccc}
0 & \longrightarrow & \check{H}^0(X,\mathcal{F}) & \longrightarrow & \check{H}^0(X,\mathcal{G}) & \longrightarrow & \check{H}^0(X,\mathcal{Q}) & \longrightarrow & \check{H}^1(X,\mathcal{F}) & \longrightarrow & 0 \\
& & \downarrow & & \downarrow & & \downarrow & & \downarrow & & \\
0 & \longrightarrow & H^0(X,\mathcal{F}) & \longrightarrow & H^0(X,\mathcal{G}) & \longrightarrow & H^0(X,\mathcal{Q}^\natural) & \longrightarrow & H^1(X,\mathcal{F}) & \longrightarrow & 0
\end{array}
$$

where the first, second and fourth vertical arrows are the morphisms (4.10). The third arrow is the morphism (4.10) in degree zero for the sheaf $\mathcal{Q}^\natural$ composed with the morphism $\check{H}^0(X,\mathcal{Q}) \to \check{H}^0(X,\mathcal{Q}^\natural)$ coming from $\mathcal{Q} \to \mathcal{Q}^\natural$.

The two leftmost vertical arrows are isomorphisms, as we already noticed. The arrow $\check{H}^0(X,\mathcal{Q}) \to H^0(X,\mathcal{Q}^\natural)$ is an isomorphism by Remark 4.4 (indeed, as we noted in Example 3.21, the natural morphism $\mathcal{Q} \to \mathcal{Q}^\natural$ is injective); as a consequence, by the Five Lemma the rightmost vertical arrow is an isomorphism as well.

Since

$$\check{H}^1(X,\mathcal{G}) = \check{H}^2(X,\mathcal{G}) = H^1(X,\mathcal{G}) = H^2(X,\mathcal{G}) = 0,$$

for $i = 2$ we can similarly draw the commutative diagram

$$
\begin{array}{ccc}
\check{H}^1(X,\mathcal{Q}) & \overset{\sim}{\longrightarrow} & \check{H}^2(X,\mathcal{F}) \\
\downarrow & & \downarrow \\
H^1(X,\mathcal{Q}^\natural) & \overset{\sim}{\longrightarrow} & H^2(X,\mathcal{F})
\end{array}
$$

Therefore, it is enough to show that the left vertical arrow in this diagram is injective. Now, first, by the previous result we can replace $H^1(X,\mathcal{Q}^\natural)$ by $\check{H}^1(X,\mathcal{Q}^\natural)$. Secondly, we consider the exact sequence

$$0 \to \mathcal{Q} \to \mathcal{Q}^\natural \to \mathcal{R} \to 0$$

---

[h]Note that in general $\check{H}^0(X,\mathcal{Q})$ is not the group of global sections of the presheaf $\mathcal{Q}$.

where by definition $\mathcal{R}$ is the quotient presheaf. Then $\check{H}^0(X, \mathcal{R}) = 0$ as $\mathcal{R}^\natural = 0$; taking the long exact sequence in Čech cohomology we obtain that $\check{H}^1(X, \mathcal{Q}) \to \check{H}^1(X, \mathcal{Q}^\natural)$ is injective.

These results about the comparison of Čech and sheaf cohomology can be strengthened. It turns out that the morphisms (4.10) are isomorphisms whenever the open cover $\mathfrak{U} = \{U_i\}$ is "fine enough" to ensure that $\mathcal{F}$ has no sheaf cohomology on all the intersections $U_{i_0\cdots i_p}$ (the so-called Leray Theorem). More precisely, we have the following theorem.

Theorem 4.33 (Leray Theorem). *Let $\mathcal{F}$ be a sheaf of abelian groups on a topological space X, and $\mathfrak{U} = \{U_i\}$ an open cover of X. Assume there exists an integer n such that*

$$H^k(U_{i_0\cdots i_p}, \mathcal{F}) = 0$$

for all k with $1 \le k \le n$ and all nonvoid intersections $U_{i_0\cdots i_p}$. Then

$$H^k(\mathfrak{U}, \mathcal{F}) \simeq H^k(X, \mathcal{F})$$

for all k in the same range.

Proof. By Remark 4.23, there exists an exact sequence

$$0 \to \mathcal{F} \to \mathcal{G} \to \mathcal{R} \to 0, \tag{4.13}$$

where $\mathcal{G}$ is flabby. The sequence

$$0 \to \mathcal{F}(U_{i_0\ldots i_p}) \to \mathcal{G}(U_{i_0\ldots i_p}) \to \mathcal{R}(U_{i_0\ldots i_p}) \to 0$$

is exact as $H^1(U_{i_0\ldots i_p}, \mathcal{F}) = 0$ by hypothesis. Hence, the sequence of complexes

$$0 \to C^\bullet(\mathfrak{U}, \mathcal{F}) \to C^\bullet(\mathfrak{U}, \mathcal{G}) \to C^\bullet(\mathfrak{U}, \mathcal{R}) \to 0$$

is exact, and therefore produces a long exact sequence for the Čech cohomologies of the open cover $\mathfrak{U}$ with coefficients in $\mathcal{F}$, $\mathcal{G}$ and $\mathcal{R}$. As $\mathcal{G}$ is flabby, $H^p(\mathfrak{U}, \mathcal{G}) = 0$ for $p > 0$ by Lemma 4.29, so that the sequence

$$0 \to H^0(\mathfrak{U}, \mathcal{F}) \to H^0(\mathfrak{U}, \mathcal{G}) \to H^0(\mathfrak{U}, \mathcal{R}) \to H^1(\mathfrak{U}, \mathcal{F}) \to 0 \tag{4.14}$$

is exact, and there are isomorphisms

$$H^p(\mathfrak{U}, \mathcal{R}) \simeq H^{p+1}(\mathfrak{U}, \mathcal{F}) \tag{4.15}$$

for $p \ge 1$. Moreover, as in Remark 4.32, comparing the sequence (4.14) with the corresponding sequence for sheaf cohomology for the exact sequence (4.13), we obtain $H^1(\mathfrak{U}, \mathcal{F}) \simeq H^1(X, \mathcal{F})$. Now we may use induction on the isomorphisms (4.15) to get the result. Note that the induction uses the fact that

$$H^k(U_{i_0\ldots i_p}, \mathcal{R}) \simeq H^{k+1}(U_{i_0\ldots i_p}, \mathcal{F}) = 0$$

for all k such that $1 \le k \le n-1$ and all p. □

Later, in Chapter 5, we shall give a proof of this theorem which uses spectral sequences (see Section 5.6.3).

Comparison with other cohomologies. In algebraic topology one attaches to a topological space X several cohomologies with coefficients in an abelian group G [29]. Loosely speaking, whenever X is paracompact and locally Euclidean, all these cohomologies coincide with the cohomology of X with coefficients in the constant sheaf G. In particular, we have the following result [63].

Proposition 4.34. *Let X be a paracompact locally Euclidean topological space. The singular cohomology of X with coefficients in a module M over a principal ideal domain is isomorphic to the sheaf cohomology of X with coefficients in the constant sheaf M_X.*

Example 4.35. The long exact sequence in cohomology associated with the exact sequence of sheaves on a complex manifold X (equation (3.1))

$$0 \to \mathbb{Z} \to \mathcal{O} \xrightarrow{\exp} \mathcal{O}^* \to 1 \tag{4.16}$$

restricted to an open set U starts with

$$0 \to H^0(U, \mathbb{Z}) \to H^0(U, \mathcal{O}) \to H^0(U, \mathcal{O}^*) \to H^1(U, \mathbb{Z}) \to \cdots.$$

This shows that the obstruction to the sequence (4.16) to be exact as a sequence of presheaves is related to the first cohomology group with coefficients in $\mathbb{Z}$. Since X, being a manifold, is paracompact and locally Euclidean, the cohomology of $\mathbb{Z}$ coincides with the singular cohomology; therefore the above-mentioned obstruction is related to the nonsimple connectedness of U.

4.4. Fine Sheaves and de Rham Cohomology

The main goal of this section is to prove a theorem stating that the de Rham cohomology of a differentiable manifold is isomorphic to the cohomology of the constant sheaf $\mathbb{R}$. To this end we shall introduce a class of sheaves of rings, the *fine sheaves*, which, more generally, are useful in the study of the cohomology of sheaves on paracompact spaces.

Definition 4.36. Given a sheaf $\mathcal{F}$ on a topological space, its support $\mathrm{Supp}(\mathcal{F})$ is defined as

$$\mathrm{Supp}(\mathcal{F}) = \{x \in X \mid \mathcal{F}_x \neq 0\}.$$

If s is a section of a sheaf $\mathcal{F}$ on some open set U, one defines

$$\mathrm{Supp}(s) = \{x \in U \mid s_x \neq 0\}.$$

$\mathrm{Supp}(s)$ is a closed subset of U, while $\mathrm{Supp}(\mathcal{F})$ need not be closed in X. A counterexample is provided by the sheaf $j_!\mathbb{Z}$, where $j : U \to X$ is the inclusion of an open nonclosed subset, and $\mathbb{Z}$ is the constant sheaf on U; in this case the support is U (the functor $j_!$ was defined in equation (4.9)).

Definition 4.37. A sheaf of rings $\mathcal{R}$ on a topological space X is *fine* if, for any locally finite open cover[i] $\mathfrak{U} = \{U_i\}_{i \in I}$ of X, there is a family $\{\rho_i\}_{i \in I}$ of global sections of $\mathcal{R}$ such that

(1) $\mathrm{Supp}(\rho_i) \subset U_i$ for all $i \in I$;
(2) $\sum_{i \in I} \rho_i = 1$.

This sum is well defined as for every point $x \in X$ only a finite number of germs $(\rho_i)_x$ are nonzero.

The family $\{\rho_i\}$ is called a *partition of unity* of the sheaf of rings $\mathcal{R}$ subordinated to the cover $\mathfrak{U}$.

For instance, the sheaf of continuous functions on a paracompact topological space as well as the sheaf of smooth functions on a differentiable manifold are fine (continuous or C^∞ partitions of unity exist),[j] while sheaves of complex or real analytic functions are not (an analytic function which is supported on a connected compact set is constant, so that analytic partitions of unity may not exist).

Proposition 4.38. *Let $\mathcal{R}$ be a fine sheaf of rings on a paracompact space X, and let $\mathcal{M}$ be a sheaf of $\mathcal{R}$-modules. Then, for every locally finite open cover $\mathfrak{U}$ of X, one has $H^q(\mathfrak{U}, \mathcal{M}) = 0$ for $q > 0$. As a consequence, $\mathcal{M}$ is acyclic, that is, $H^q(X, \mathcal{M}) = 0$ for $q > 0$.*

---

[i] See footnote a in this chapter.
[j] References are [15] for paracompact topological spaces, and [7, 63] for C^∞ partitions of unity.

Proof. Let $\mathfrak{U} = \{U_i\}_{i \in I}$ be a locally finite open cover of X, and let $\{\rho_i\}$ be a partition of unity of $\mathcal{R}$ subordinated to $\mathfrak{U}$. For any $\alpha \in C^q(\mathfrak{U}, \mathcal{M})$ with $q > 0$ we set

$$(K\alpha)_{i_0 \cdots i_{q-1}} = \sum_{k=0}^{q} (-1)^k \sum_{\substack{j \in I \\ i_{k-1} < j < i_k}} \rho_j \, \alpha_{i_0 \cdots i_{k-1} j i_k \cdots i_{q-1}}.$$

This defines a morphism $K : C^q(\mathfrak{U}, \mathcal{M}) \to C^{q-1}(\mathfrak{U}, \mathcal{M})$ such that $\delta K + K\delta = \mathrm{id}$ (i.e., K is a null homotopy operator); then $\alpha = \delta K \alpha$ if $\delta \alpha = 0$, so that $H^q(\mathfrak{U}, \mathcal{M}) = 0$ for $q > 0$. Since on a paracompact space the locally finite open covers are cofinal in the family of all covers, we can take direct limit on such covers, thus getting $\check{H}^q(X, \mathcal{M}) = 0$ for $q > 0$. By Theorem 4.31, also $H^q(X, \mathcal{M}) = 0$. $\qquad\square$

Let $(\Omega_X^\bullet(X), d)$ be the de Rham complex over a differentiable manifold X, where the $\Omega_X^\bullet(X)$ are the spaces of global differential forms on X, i.e., the global sections of the sheaves $\Omega_X^\bullet$. The cohomology groups of this complex are denoted $H_{dR}^\bullet(X)$ and are called the *de Rham cohomology groups* of X (these were already introduced in Example 3.13). Note that the sheaves $\Omega_X^\bullet$ are $\mathcal{C}_X^\infty$-modules, where $\mathcal{C}_X^\infty$ is the sheaf of C^∞ functions on X.

Example 4.39 (The Mayer–Vietoris sequence). Given a differentiable manifold X, let $\mathfrak{U}$ be an open cover formed by two sets U and V. Since $C^2(\mathfrak{U}, \Omega_X^k) = 0$ (there are no triple intersections) we have a sequence

$$0 \to H^0(\mathfrak{U}, \Omega_X^k) \to C^0(\mathfrak{U}, \Omega_X^k) \overset{\delta}{\to} C^1(\mathfrak{U}, \Omega_X^k) \to 0$$

which a priori is exact everywhere but at $C^1(\mathfrak{U}, \Omega_X^k)$; the cokernel of δ is the cohomology group $H^1(\mathfrak{U}, \Omega_X^k)$. However, by Proposition 4.38 one has $H^1(\mathfrak{U}, \Omega_X^k) = 0$, which means that δ is surjective, and the sequence is exact at that place as well (one can also directly prove that δ is surjective by using a partition of unity argument). We have the identifications

$$H^0(\mathfrak{U}, \Omega_X^k) = \Omega_X^k(X), \quad C^0(\mathfrak{U}, \Omega_X^k) = \Omega_X^k(U) \oplus \Omega_X^k(V),$$

$$C^1(\mathfrak{U}, \Omega_X^k) = \Omega_X^k(U \cap V)$$

so that we obtain the exactness of the sequence

$$0 \to \Omega_X^k(X) \to \Omega_X^k(U) \oplus \Omega_X^k(V) \to \Omega_X^k(U \cap V) \to 0.$$

If we look at this as an exact sequence of complexes and take cohomology we obtain the long exact sequence

$$0 \to H_{dR}^0(X) \to H_{dR}^0(U) \oplus H_{dR}^0(V) \to H_{dR}^0(U \cap V)$$
$$\to H_{dR}^1(X) \to H_{dR}^1(U) \oplus H_{dR}^1(V) \to H_{dR}^1(U \cap V) \to H_{dR}^2(X) \to \cdots,$$

which is called the *Mayer–Vietoris sequence*.

The abstract de Rham Theorem 2.23 implies that the de Rham cohomology of X is isomorphic to the sheaf cohomology of the constant sheaf $\mathbb{R}$. This relies on the following fact, called *Poincaré Lemma* [7].

Lemma 4.40 (Poincaré Lemma). *The cohomology groups $H_{dR}^k(\mathbb{R}^n)$ are zero for $n \geq 1$ and $k \geq 1$.*

Proof. A possible proof is obtained by writing a null homotopy for the de Rham complex $(\Omega_{\mathbb{R}^n}^\bullet(\mathbb{R}^n), d)$. If ω is a k-form on $\mathbb{R}^n$, define

$$(K\omega)(x)$$

$$= \sum_{i_1 < \cdots < i_k} \sum_{p=1}^{k} (-1)^{p-1} \left[\int_0^1 t^{k-1} \omega_{i_1 \cdots i_k}(tx) dt \right] x^{i_p} dx^{i_1} \wedge \cdots \widehat{dx^{i_p}} \cdots \wedge dx^{i_k}.$$

Using integration by parts, one can prove that K satisfies the condition $K \circ d + d \circ K = \mathrm{id}$. $\qquad \square$

As a consequence, since every point of an n-dimensional differentiable manifold X has a neighbourhood which is diffeomorphic to $\mathbb{R}^n$, the *de Rham sheaf complex*

$$\Omega_X^0 \xrightarrow{d} \Omega_X^1 \xrightarrow{d} \Omega_X^2 \xrightarrow{d} \cdots$$

(where Ω_X^0 is another name for $\mathcal{C}_X^\infty$) is exact in positive degree. Moreover, the kernel of $d : \Omega_X^0 \to \Omega_X^1$ is the constant sheaf $\mathbb{R}$, so that the de Rham sheaf complex is a resolution of $\mathbb{R}$. Furthermore, the resolution is acyclic by Proposition 4.38. Therefore, we obtain the following theorem.

Theorem 4.41 (de Rham Theorem). *Let X be a differentiable manifold. For all $k \geq 0$, the cohomology groups $H_{dR}^k(X)$ and $H^k(X, \mathbb{R})$ are isomorphic.*

Good covers. By means of the Leray Theorem 4.33 we may reduce the problem of computing the Čech cohomology of a differentiable manifold with coefficients in the constant sheaf $\mathbb{R}$ (which, via the de Rham Theorem, amounts to computing its de Rham cohomology) to the computation of the cohomology of a cover with coefficients in $\mathbb{R}$; thus a problem which in principle would need the solution of differential equations on topologically nontrivial manifolds is reduced to a simpler problem which only involves the intersection pattern of the open sets of a cover.

Definition 4.42. An open cover $\mathfrak{U}$ of a differentiable manifold is *good* if all nonempty intersections of a finite number of its members are diffeomorphic to $\mathbb{R}^n$.

Good covers exist on any differentiable manifold (cf. [40]). Due to Poincaré Lemma, the constant sheaf $\mathbb{R}$ is acyclic on all intersections of open sets in a good cover. We have therefore:

Proposition 4.43. *For any good cover $\mathfrak{U}$ of a differentiable manifold X, one has isomorphisms*

$$H^k(\mathfrak{U}, \mathbb{R}) \simeq H^k(X, \mathbb{R}), \quad k \geq 0.$$

So by the de Rham Theorem 4.41 we also have $H^k(\mathfrak{U}, \mathbb{R}) \simeq H^k_{dR}(X)$. The cover of Example 4.2 was good, so we computed there the de Rham cohomology of the circle S^1.

Dolbeault Theorem. The Dolbeault Theorem is a holomorphic analogue of the de Rham Theorem. Let X be an n-dimensional complex manifold. For every nonnegative p, q, let $\Omega_X^{p,q}$ be the sheaf of differential forms of type (p, q), i.e., the sheaf of complex-valued $(p+q)$-differential forms that locally can be written as

$$\eta = \sum_{\substack{i_1,\ldots,i_p=1,\ldots,n \\ j_1,\ldots,j_q=1,\ldots,n}} \eta_{i_1,\ldots,i_p,j_1,\ldots,j_q} dz^{i_1} \wedge \cdots \wedge dz^{i_p} \wedge d\bar{z}^{j_1} \wedge \cdots \wedge d\bar{z}^{j_q},$$

where $(z^1, \ldots, z^n)$ are local holomorphic coordinates, and $\eta_{i_1,\ldots,i_p,j_1,\ldots,j_q}$ are C^∞ functions. The Cauchy–Riemann operator $\bar{\partial}$ is a $\mathbb{C}$-linear morphism $\Omega_X^{p,q} \to \Omega_X^{p,q+1}$, and squares to zero, so that for every p one has a Dolbeault complex $(\Omega_X^{p,\bullet}, \bar{\partial})$ [21]. The Dolbeault cohomology groups are defined as

$$H_{\bar{\partial}}^{p,q}(X) = H^q(\Omega_X^{p,\bullet}(X), \bar{\partial}).$$

Moreover, the following facts hold true:

- The operator $\bar{\partial}$ satisfies a Poincaré Lemma, i.e., $\bar{\partial}$-closed differential forms are locally $\bar{\partial}$-exact.
- There is an inclusion $\Omega_X^p \hookrightarrow \Omega_X^{p,0}$, whose image is the kernel of $\bar{\partial} : \Omega_X^{p,0} \to \Omega_X^{p,1}$.
- The sheaves $\Omega_X^{p,q}$ are $\mathcal{C}_X^\infty$-modules, so that by Proposition 4.38 they are acyclic.

Therefore, the Dolbeault complex yields acyclic resolutions of the sheaves of holomorphic forms:

$$0 \to \Omega_X^p \to \Omega_X^{p,0} \xrightarrow{\bar{\partial}} \Omega_X^{p,1} \xrightarrow{\bar{\partial}} \cdots . \tag{4.17}$$

Now, the abstract de Rham Theorem 2.23 implies the Dolbeault Theorem.

Theorem 4.44. *For every $p \geq 0$, the groups $H^q(X, \Omega_X^p)$ and $H_{\bar{\partial}}^{p,q}(X)$ are isomorphic.*

4.5. Sheaf Cohomology of Schemes

In this section we shall give some basic results about the cohomology of a special class of sheaves (the quasi-coherent sheaves) on a scheme. Schemes were introduced by Grothendieck [23], who subsequently developed their theory [24–26], as a powerful generalization of the notion of variety, allowing for the presence of nilpotent "functions". The notion of scheme stresses the deep algebraic content of algebraic geometry, allowing for the application of powerful algebraic techniques. A more modern and pedagogical introduction to schemes is given in Hartshorne's book [28].

The regular functions on a scheme are obtained, at least locally, by applying an algebraic procedure called "localization" to a ring; similarly, starting from modules over these rings one obtains a class of sheaves named *quasi-coherent sheaves*. The cohomology of quasi-coherent sheaves on schemes satisfying suitable conditions is another instance where Čech and sheaf cohomology coincide. The aim of this section is to prove this isomorphism. To this end, we shall review the basic definitions and constructions in the theory of schemes.

The spectrum of a ring. The building blocks of scheme theory are the *affine schemes*, which are the geometric counterparts of rings. Let R be a commutative ring with unity. As we already mentioned in Example 3.25, the *prime spectrum of* R, denoted $\operatorname{Spec} R$, is the set of prime ideals of R,[k] equipped with a topology (the Zariski topology) whose closed sets are of the form

$$V(\mathfrak{r}) = \{\text{all prime ideals } \mathfrak{p} \text{ of } R \text{ containing } \mathfrak{r}\},$$

where $\mathfrak{r}$ is an ideal of R. It is not difficult to check that this defines indeed a topology for $\operatorname{Spec} R$.

Example 4.45. If R is a field $\Bbbk$, then $\operatorname{Spec} \Bbbk$ has only one point, the zero ideal. This topological space is trivially Hausdorff.

Example 4.46. Let $R = \Bbbk[x]$, the ring of polynomials in one variable with coefficients in a field $\Bbbk$. Its spectrum is denoted by $\mathbb{A}^1_{\Bbbk}$ and is called *the affine line over* $\Bbbk$. It has a point, the zero ideal, which is dense in the whole space (this is called the *generic point*). So $\mathbb{A}^1_{\Bbbk}$ is not a T1 space. When $\Bbbk$ is algebraically closed, the closed points of $\mathbb{A}^1_{\Bbbk}$ are in a one-to-one correspondence with the elements of $\Bbbk$.

Example 4.47. In general the spectrum of a ring has more nonclosed points than the zero ideal. The *height* of a prime ideal $\mathfrak{p}$ in R is the supremum of the natural numbers n for which there is a chain of inclusions of distinct prime ideals

$$\mathfrak{p}_0 \subset \mathfrak{p}_1 \subset \cdots \subset \mathfrak{p}_n = \mathfrak{p}.$$

The *(Krull) dimension of* R is the supremum of the heights of all prime ideals in R. Assume that R has Krull dimension ≥ 2.[1] Then there exists at least one prime ideal $\mathfrak{p}$ with a chain of distinct prime ideals $\mathfrak{p}_0 \subset \mathfrak{p}_1 \subset \mathfrak{p}$ (with $\mathfrak{p}_1 \neq 0$) so that $\mathfrak{p}$ is in the closure $V(\mathfrak{p}_1)$ of $\mathfrak{p}_1$, and $\mathfrak{p}_1$ is not closed.

Example 4.48. Let $R = \Bbbk[x, y]$ be the polynomial ring in two variables over an algebraically closed field $\Bbbk$. Then $\operatorname{Spec} R$ is the affine plane $\mathbb{A}^2_{\Bbbk}$. Its closed points are in a one-to-one correspondence with the elements of $\Bbbk^2$. Moreover, it has the generic point, corresponding to the zero ideal, which is dense in $\mathbb{A}^2_{\Bbbk}$. If P is an irreducible polynomial, the ideal I_P generated by P is prime, so that P corresponds to a (nonclosed) point η in $\mathbb{A}^2_{\Bbbk}$; its closure $V(I_P)$ contains η and the closed points $(x, y) \in \mathbb{A}^2_{\Bbbk}$ such that $P(x, y) = 0$. Therefore, η is the *generic point* of the curve $P = 0$.

---

[k]A proper ideal $\mathfrak{p}$ of R is prime if, given $x, y \in R$, the condition $xy \in \mathfrak{p}$ implies that at least one of x and y are in $\mathfrak{p}$.

[1]Such rings exist, for instance, $\Bbbk[x_1, \ldots, x_n]$ has Krull dimension n [46, Chapter 2].

The next step is to equip the spectrum $\operatorname{Spec} R$ with a sheaf of functions. This is done via a procedure called *localization*. A *multiplicative system* S in R is a subset which contains the unity and is closed under multiplication. The *localization of R at S*, denoted R_S, is the quotient of $R \times S$ by the equivalence relation

$$(r, s) \sim (r', s') \quad \text{if there exists } t \in S \text{ such that } t(rs' - r's) = 0.$$

The equivalence class of a pair (r, s) is usually denoted r/s; this reflects the intuitive idea that the elements of S become invertible in R_S. R_S inherits a structure of commutative ring with unity.

Example 4.49. If R is an integral domain, $S = R - \{0\}$ is a multiplicative system, and R_S is a field (the *field of fractions* of R).

Example 4.50. If $\mathfrak{p} \subset R$ is a prime ideal, then $S = R - \mathfrak{p}$ is a multiplicative system. The localized ring is denoted by $R_\mathfrak{p}$. It is a *local ring,* i.e., it has a unique maximal ideal $\mathfrak{m}_\mathfrak{p}$, given by the elements r/s, with $r \in \mathfrak{p}$ and $s \in S$.

Example 4.51. For a given $f \in R$, let $S = \{f^n\}_{n \in \mathbb{N}}$. The localized ring is denoted by R_f, and its spectrum by $D(f)$. It is an open subset of $\operatorname{Spec} R$ (the complement of the closed subset $V((f))$, i.e., the closed subset corresponding to the ideal generated by f), and indeed such subsets, with varying $f \in R$, are a basis of open sets for the topology of $\operatorname{Spec} R$ (intuitively, this is the open set where f is multiplicatively invertible).

A sheaf $\mathcal{O}$ on $\operatorname{Spec} R$ is defined by associating to any open subset $U \subset \operatorname{Spec} R$ the set $\mathcal{O}(U)$ of maps $s : U \to \prod_{\mathfrak{p} \in U} R_\mathfrak{p}$ such that:

- $s(\mathfrak{p}) \in R_\mathfrak{p}$ for all $\mathfrak{p} \in U$;
- s is locally a quotient of elements of R, that is, for each $\mathfrak{p} \in U$ there exist an open neighbourhood V of $\mathfrak{p}$ and elements $f, g \in R$, such that for each $\mathfrak{q} \in V$, one has $g \notin \mathfrak{q}$, and $s(\mathfrak{q}) = f/g$ in $R_\mathfrak{q}$.

$\mathcal{O}(U)$ is clearly a commutative ring with unity, and restriction morphisms are defined just by restricting the maps $s \in \mathcal{O}(U)$ to open subsets. This defines a sheaf of rings $\mathcal{O}$ on $\operatorname{Spec} R$. We list some properties of this sheaf:

- by definition, $\mathcal{O}_\mathfrak{p} \simeq R_\mathfrak{p}$ for every point $\mathfrak{p} \in \operatorname{Spec} R$;
- on the open sets of the form $D(f)$ one has $\mathcal{O}(D(f)) \simeq R_f$ (this provides a more intuitive grasp of the definition of $\mathcal{O}$);
- in particular, $\Gamma(\operatorname{Spec} R, \mathcal{O}) \simeq R$.

The stalk $\mathcal{O}_\mathfrak{p}$ of $\mathcal{O}$ at $\mathfrak{p}$ is usually called the *local ring* of $\mathfrak{p}$; it is a local ring (see Example 4.50) and the quotient field $\mathcal{O}_\mathfrak{p}/\mathfrak{m}_\mathfrak{p}$ is called the *residue field* at $\mathfrak{p}$.

Example 4.52. The topological space $\operatorname{Spec}\mathbb{Z}$ contains the generic point, corresponding to the zero ideal, whose local ring is $\mathbb{Q}$; all other points are closed and are in a one-to-one correspondence with the prime natural numbers. This provides an example where the morphism $\left(\prod_{i\in I}\mathcal{O}_X\right)_x \to \prod_{i\in I}\mathcal{O}_{X,x}$ in Example 3.25 is not an isomorphism. For $R = \mathbb{Z}$, and x the generic point, the morphism takes the form

$$\left(\prod_{i\in I}\mathbb{Z}\right)\otimes_\mathbb{Z}\mathbb{Q} \to \prod_{i\in I}\mathbb{Q}$$

given by $(z_i)\otimes q \mapsto (z_i q)$. This morphism is not surjective.

Schemes. We introduced ringed spaces in Definition 4.9. We refine that notion in the following way.

Definition 4.53. A locally ringed space $(X,\mathcal{O}_X)$ is a ringed space (see Definition 4.9) such that for each $x \in X$, the stalk $\mathcal{O}_{X,x}$ is a local ring, i.e., a ring with a unique maximal ideal. A morphism between two locally ringed spaces $(X,\mathcal{O}_X)$ and $(X',\mathcal{O}_{X'})$ is a morphism of ringed spaces $(f,f^\sharp)$ such that for every $x \in X$, the induced morphism between the stalks $\mathcal{O}_{X',f(x)} \to \mathcal{O}_{X,x}$ is a morphism of local rings (i.e., the inverse image of the maximal ideal of $\mathcal{O}_{X,x}$ is the maximal ideal of $\mathcal{O}_{X',f(x)}$).

As we noted in Example 4.50, for every $\mathfrak{p} \in \operatorname{Spec} R$, the ring $R_\mathfrak{p}$ is local, so that $(\operatorname{Spec} R, \mathcal{O})$ is a locally ringed space.

Now we define *affine schemes* and *schemes*.

Definition 4.54. An affine scheme is a locally ringed space $(X,\mathcal{O}_X)$ isomorphic to the spectrum of a ring R. A scheme $(X,\mathcal{O}_X)$ is a locally ringed space where every point $x \in X$ has an open neighbourhood U such that $(U,\mathcal{O}_X|_U)$ is an affine scheme.

The cohomology of the class of sheaves we are going to consider is well behaved for schemes that satisfy a kind of finiteness condition, i.e., they are *noetherian*. One starts from a purely algebraic notion: a commutative ring is said to be noetherian if it satisfies the *ascending chain condition for ideals,* i.e., any increasing chain of nested ideals

$$\mathfrak{r}_0 \subset \mathfrak{r}_1 \subset \mathfrak{r}_2 \subset \cdots$$

stabilizes after a finite number of steps.[m]

Definition 4.55. A scheme X is locally noetherian if there is an open cover $\mathfrak{U} = \{U_i\}$ where each open U_i is the spectrum of a noetherian ring. If the cover can be taken to be finite, X is said to be a noetherian scheme.

Another requisite for having a good cohomology theory is that the scheme satisfies some kind of separatedness condition. This cannot be the Hausdorff condition. Indeed, a topological space X is Hausdorff if and only if the diagonal $\Delta \subset X \times X$ is a closed subset of $X \times X$ equipped with the product topology [37]. However, the product topology is not a good topology in algebraic geometry. For instance, the affine plane $\mathbb{A}_\Bbbk^2 = \operatorname{Spec} \Bbbk[x, y]$, as a topological space, is not the cartesian product $\mathbb{A}_\Bbbk^1 \times \mathbb{A}_\Bbbk^1$ equipped with the product topology, as the reader can easily check (think what are the closed sets in $\mathbb{A}_\Bbbk^1 \dots$).

What one needs to do is to give a good definition of product of schemes. The guiding principle is that the product should satisfy a universal property.

We will say that a scheme over S (a fixed scheme) is a pair (X, f), where X is a scheme and $f : X \to S$ is a morphism of schemes. Usually one says things like "let X be a scheme over S", understanding the morphism f. Since every scheme morphism locally is induced by a homomorphism of rings, and there is a unique ring morphism from $\mathbb{Z}$ to any commutative ring with unity, all schemes X are schemes over $\mathbb{Z}$. When no base scheme S is specified, we will think of X as a scheme over $\mathbb{Z}$.

Let X, Y be two schemes over S; the fibred product $X \times_S Y$ is a scheme over S with morphisms $X \times_S Y \to X$, $X \times_S Y \to Y$ satisfying the following properties:

- the diagram

$$
\begin{array}{ccc}
X \times_S Y & \longrightarrow & Y \\
\downarrow & \searrow & \downarrow \\
X & \longrightarrow & S
\end{array}
$$

commutes;

---

[m]The topological counterpart of this notion is that of *noetherian topological space*, which satisfies a *descending* chain condition for closed subset.

- (universality) for every commutative diagram of morphisms of schemes over S

$$\begin{array}{ccc} Z & \longrightarrow & Y \\ \downarrow & & \downarrow \\ X & \longrightarrow & S \end{array}$$

there exists a unique morphism $Z \to X \times_S Y$ such that the following diagram

$$\begin{array}{ccc} Z & & \\ & X \times_S Y & \longrightarrow & Y \\ & \downarrow & & \downarrow \\ & X & \longrightarrow & S \end{array}$$

commutes.

If $X = \operatorname{Spec} A$, $Y = \operatorname{Spec} B$ and $S = \operatorname{Spec} C$ are affine schemes with ring morphisms $C \to A$ and $C \to B$ (inducing morphism $X \to S$ and $Y \to S$), the fibred product $X \times_S Y$ may be defined as $\operatorname{Spec}(A \otimes_C B)$, with morphisms $X \times_S Y \to X$ and $X \times_S Y \to Y$ given by the natural morphisms $A \to A \otimes_C B$ and $B \to A \otimes_C B$.

In the nonaffine case, the fibred product is locally defined as above and then glued. See [28, Theorem II.3.3] for details.

We shall need the notion of a closed immersion.

Definition 4.56. A scheme morphism $f : Y \to X$ is said to be a closed immersion if it induces a homeomorphism onto a closed subset of the topological space X and the induced morphism of sheaves on X, $f^\sharp : \mathcal{O}_X \to f_* \mathcal{O}_Y$, is surjective.

Example 4.57. If R is a commutative ring with unity, and $\mathfrak{r}$ is an ideal, let $X = \operatorname{Spec} R$ and $Y = \operatorname{Spec} R/\mathfrak{r}$. The projection of rings $R \to R/\mathfrak{r}$ induces a morphism of schemes $i : Y \to X$; this is a closed immersion. Indeed i provides a homeomorphism $Y \to V(\mathfrak{r}) \subset X$. Moreover, given $\mathfrak{p} \in X$, the projection $\pi : R \to R/\mathfrak{r}$ yields a ring epimorphism between the localizations $R_\mathfrak{p} \to (R/\mathfrak{r})_{\pi(\mathfrak{p})}$, which means that the morphism $\mathcal{O}_X \to i_* \mathcal{O}_Y$ is surjective.

Now we can give the definition of separated scheme. Given a scheme X over S, the *diagonal morphism* is the morphism $\Delta : X \to X \times_S X$ defined by the diagram

$$(4.18)$$

Definition 4.58. A scheme morphism $f : X \to Y$ is separated if the diagonal morphism $\Delta : X \to X \times_Y X$ is a closed immersion. In particular, the scheme X is separated if the canonical morphism $X \to \operatorname{Spec} \mathbb{Z}$ is separated.

Remark 4.59. A scheme X over a field $\Bbbk$ is a scheme over $\operatorname{Spec} \Bbbk$. A variety over $\Bbbk$ is a reduced,[n] separated scheme of finite type[o] over $\Bbbk$. By Corollary II.4.6(e) in [28], a reduced scheme of finite type over $\Bbbk$ which is separated over $\mathbb{Z}$ is a variety.

All affine schemes are separated. Indeed, if $X = \operatorname{Spec} R$ is the spectrum of a ring, the fibred product of locally ringed spaces

$$\operatorname{Spec} R \times_{\operatorname{Spec} \mathbb{Z}} \operatorname{Spec} R = \operatorname{Spec}(R \otimes_{\mathbb{Z}} R)$$

is given by the homomorphism of rings $\mathbb{Z} \to R$. The diagonal morphism $\Delta : \operatorname{Spec} R \to \operatorname{Spec}(R \otimes_{\mathbb{Z}} R)$ is given by the surjective ring morphism (multiplication morphism) $\rho : R \otimes_{\mathbb{Z}} R \to R$, i.e., $f \otimes f' \to ff'$. Then apply Example 4.57 with R replaced by $R \otimes_{\mathbb{Z}} R$, and $\mathfrak{r} = \ker \rho$. So Δ is a closed immersion, and the spectrum of any ring is a separated scheme.

The separatedness of a nonaffine scheme can be checked by means of a simple criterion; indeed, a scheme is separated if and only if the image of the diagonal morphism is closed. As one sees in the diagram (4.18), the diagonal morphism factorizes the identity morphism $X \to X$, so that $\Delta : X \to \Delta(X)$ is a homeomorphism provided that $\Delta(X)$ is a closed subset. The surjectivity of the morphism $\mathcal{O}_{X \times_{\operatorname{Spec} \mathbb{Z}} X} \to \Delta_* \mathcal{O}_X$ is checked locally, choosing an open

---

[n]Reduced means that the structure sheaf has no nilpotents.
[o]A scheme over a field $\Bbbk$ is of finite type if it has a finite affine open cover $\{U_i\}$, where each U_i is the spectrum of a finitely generated $\Bbbk$-algebra. Note that an algebraic variety is always noetherian as a finitely generated $\Bbbk$-algebra is noetherian.

affine neighbourhood U of a given point $x \in X$. Then, $U \times_{\mathrm{Spec}\,\mathbb{Z}} U$ is an open affine subset of $\Delta(X)$. But $\Delta_{|U} : U \to U \times_{\mathrm{Spec}\,\mathbb{Z}} U$ is a closed immersion (as all affine schemes are separated), so that $\mathcal{O}_{U \times_{\mathrm{Spec}\,\mathbb{Z}} U} \to \Delta_* \mathcal{O}_U$ is surjective.

So we have proved the following.

Proposition 4.60. *A scheme X is separated if and only if $\Delta(X)$ is a closed subset of $X \times_{\mathrm{Spec}\,\mathbb{Z}} X$.*

Projective schemes. An interesting class of schemes is that of projective schemes, which are not affine, and are obtained by a construction (the projective spectrum) whose starting point is a *graded ring*. By that one means a ring R, with we assume to be commutative with unity, with a direct sum decomposition into abelian groups R_i:

$$R = \bigoplus_{i \in \mathbb{N}} R_i$$

with the condition $R_i R_j \subset R_{i+j}$. An element of R is said to be *homogeneous* if it lies in one of the summands R_i. A *graded ideal* of R is an ideal which is generated by homogeneous elements. Finally, R_+ is the ideal

$$R_+ = \bigoplus_{i > 0} R_i.$$

Definition 4.61.

- Proj R is the set of all homogeneous prime ideals of R that do not contain all of R_+.
- For every homogeneous ideal $\mathfrak{a}$ of R let

$$V(\mathfrak{a}) = \{\mathfrak{p} \in \mathrm{Proj}\,R \,|\, \mathfrak{a} \subset \mathfrak{p}\}.$$

- the set Proj R is topologized by declaring that its closed subsets have the form $V(\mathfrak{a})$ for some homogeneous ideal $\mathfrak{a}$.

A sheaf of rings $\mathcal{O}$ is defined on Proj R by suitably modifying the construction of the sheaf of regular functions on the spectrum of a ring; the pair $(\mathrm{Proj}\,R, \mathcal{O})$ is a scheme. For details, see e.g. [28, Section II.2].

Example 4.62. Let R be the ring of polynomials in $n + 1$ variables with coefficients in a commutative ring with unity S:

$$R = S[x_0, \ldots, x_n],$$

whose homogeneous summand of degree i is the module of homogeneous polynomials of degree i. Then $\operatorname{Proj} R$ is the n-dimensional projective space over S, denoted $\mathbb{P}_S^n$. Since S injects into $S[x_0, \ldots, x_n]$, there is a morphism $\mathbb{P}_S^n \to \operatorname{Spec} S$, i.e., $\mathbb{P}_S^n$ is a scheme over $\operatorname{Spec} S$.

If S is a field $\Bbbk$, $\mathbb{P}_{\Bbbk}^n$ is a scheme over $\Bbbk$. If $\Bbbk$ is algebraically closed, the set of closed points of $\mathbb{P}_{\Bbbk}^n$ is the quotient

$$\frac{\Bbbk^{n+1} - \{0\}}{\Bbbk^*},$$

where $\Bbbk^*$ acts by multiplying the entries of $\Bbbk^{n+1}$. In particular, when $\Bbbk = \mathbb{C}$ this is the complex projective n-space.

The n-dimensional projective space $\mathbb{P}_{\Bbbk}^n$ is of finite type (see footnotes n and o) because it is covered by $n+1$ open sets that are isomorphic to the n-dimensional affine space $\mathbb{A}_{\Bbbk}^n = \operatorname{Spec} \Bbbk[x_1, \ldots, x_n]$.

Quasi-coherent sheaves. Now we recall the notion of sheaf of modules, which we have already introduced in Section 4.2. A *sheaf of $\mathcal{O}_X$-modules* over a ringed space $(X, \mathcal{O}_X)$ is a sheaf of groups $\mathcal{F}$ over X together with an assignment, for each open set $U \subset X$, of an $\mathcal{O}_X(U)$-module structure on the group $\mathcal{F}(U)$, compatibly with the restriction morphisms $\mathcal{F}(U) \to \mathcal{F}(V)$, where $\mathcal{F}(V)$ has an $\mathcal{O}_X(U)$-module structure via the morphism $\mathcal{O}_X(U) \to \mathcal{O}_X(V)$.

We also know that the category $\mathcal{O}_X$-**mod** of sheaves of $\mathcal{O}_X$-modules over a ringed space $(X, \mathcal{O}_X)$ is an abelian category with enough injectives, see Proposition 4.10.

The *sheaf associated* to a module M over a ring R is a sheaf of $\mathcal{O}_{\operatorname{Spec} R}$-modules over $\operatorname{Spec} R$. We denote it as $\tilde{M}$. This assignment is given as follows: for each point $x \in \operatorname{Spec} R$, let M_x be the localization of M at x,[p] and for each open set U in $\operatorname{Spec} R$, consider the set of functions

$$s : U \to \coprod_{x \in U} M_x$$

such that $s(x) \in M_x$, and s is locally a quotient m/r, where $m \in M$ and $r \in R$. This means that for each $x \in U$ there is a neighbourhood $V \subset U$ containing x and elements $m \in M$ and $r \in R$ such that for each $y \in V$,

---

[p]The localization of an R-module M is defined in analogy with that of a ring, see for instance [1, 46].

$r \notin y$ and $s(y) = m/r \in M_y$. The obvious restriction morphisms make $\tilde{M}$ into a sheaf. Moreover, the correspondence $M \rightsquigarrow \tilde{M}$ is an exact functor from the category of R-modules to the category of $\mathcal{O}_{\operatorname{Spec} R}$-modules, and $\Gamma(\operatorname{Spec} R, \tilde{M}) \simeq M$.

As the open subsets $D(f)$, for $f \in R$, form a basis for the Zariski topology of $\operatorname{Spec} R$, one could define $\tilde{M}$ by letting $\tilde{M}(D(f)) = M_f$ as R_f-modules, and for any inclusion $D(f) \subset D(g)$ the restriction morphisms given by the natural morphisms $\rho_{g,f} : M_g \to M_f$.

We give here the definitions of quasi-coherent and coherent sheaf.

Definition 4.63. A sheaf of $\mathcal{O}_X$-modules $\mathcal{F}$ over a scheme $(X, \mathcal{O}_X)$ is quasi-coherent if there exist an affine open cover $(U_i = \operatorname{Spec} R_i)_{i \in I}$ of X and R_i-modules M_i such that $\mathcal{F}|_{U_i} \simeq \tilde{M}_i$ for every $i \in I$. If each M_i is finitely generated over R_i, $\mathcal{F}$ is said to be coherent.

Moreover, it can be proved (see [28, Proposition II.5.4] for a detailed proof) that a sheaf $\mathcal{F}$ of $\mathcal{O}_X$-modules is quasi-coherent if and only if for any affine open subset $U = \operatorname{Spec} R$ of X there exists an R-module M such that $\mathcal{F}|_U \simeq \tilde{M}$. If X is noetherian, $\mathcal{F}$ is coherent if and only if for any affine open subset of X the corresponding R-module is finitely generated.

When $X = \operatorname{Spec} R$, the functor

$$\sim : R\text{-}\mathbf{mod} \to \mathcal{O}_{\operatorname{Spec} R}\text{-}\mathbf{mod}$$

gives rise to an equivalence of the category R-**mod** with the category of quasi-coherent $\mathcal{O}_X$-modules, whose inverse is the global section functor Γ. Moreover, if R is noetherian, there is also an equivalence between the categories of finitely generated R-modules and of coherent $\mathcal{O}_X$-modules. For more details see [28, Section II.5].

Cohomology of quasi-coherent sheaves. Finally, we can study the cohomology of quasi-coherent sheaves on schemes. One might hope that the functor $\sim : R$-**mod** $\to \mathcal{O}_{\operatorname{Spec} R}$-**mod** maps injectives to injectives; actually only a weaker result holds, which however will be sufficient to our purposes.

Proposition 4.64. *Let R be a noetherian ring and I an injective R-module. The sheaf $\tilde{I}$ on $X = \operatorname{Spec} R$ is flabby.*

The proof of this proposition requires Krull's Theorem and two Lemmas. We only state the part of Krull's Theorem that will be needed.

Theorem 4.65. *Let R be a noetherian ring, let $\mathfrak{r}$ be an ideal and let M and N be finitely generated R-modules, with $M \subset N$. For every $n > 0$, there exists $m \geq n$ such that $\mathfrak{r}^m N \cap M \subset \mathfrak{r}^n M$.*

For the proof of Proposition 4.64 and of the required lemmas we shall follow [28, Section III.3].

Lemma 4.66. *Let $\mathfrak{r}$ be an ideal of a noetherian ring R, and let I be an injective R-module. The submodule*

$$\Gamma_\mathfrak{r}(I) = \{x \in I \mid \mathfrak{r}^n x = 0 \text{ for some } n > 0\}$$

is injective.

Proof. Denote $J = \Gamma_\mathfrak{r}(I)$. By Baer's Criterion (see Appendix A.3) we only need to show that every morphism $\phi : \mathfrak{s} \to J$, where $\mathfrak{s}$ is an ideal of R, extends to R. Since R is noetherian, $\mathfrak{s}$ is finitely generated, and one can apply Krull's Theorem 4.65, taking $M = \mathfrak{s}$ and $N = R$. Since $\mathfrak{s}$ is finitely generated, there exists $n > 0$ such that $\mathfrak{r}^n \phi(b) = 0$ for all $b \in \mathfrak{s}$; then there is $m \geq n$ such that $\mathfrak{s} \cap \mathfrak{r}^m \subset \mathfrak{r}^n \mathfrak{s}$, so that ϕ factors through $\mathfrak{s}/\mathfrak{s} \cap \mathfrak{r}^m$ (as $\phi(\mathfrak{s} \cap \mathfrak{r}^m) = 0$). From the diagram

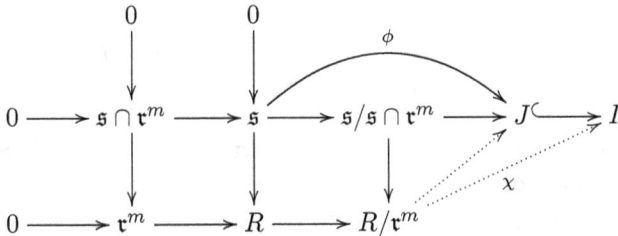

the morphism $\mathfrak{s}/\mathfrak{s} \cap \mathfrak{r}^m \to J \hookrightarrow I$ can be extended (I is injective) to $\chi : R/\mathfrak{r}^m \to I$, whose image is actually contained in J. Therefore, the composition $R \to R/\mathfrak{r}^n \to J$ is an extension of ϕ to R. $\square$

Lemma 4.67. *If I is an injective module over a noetherian ring R, and $f \in R$, the natural morphism $I \to I_f$ is surjective.*

Proof. Denote by $\mathfrak{q}_k$ the annihilator of f^k in R. As R is noetherian, the chain $\mathfrak{q}_1 \subset \mathfrak{q}_2 \subset \cdots$ stabilizes, say at step r, i.e., $\mathfrak{q}_r = \mathfrak{q}_{r+1} = \cdots$. Any element $a \in I_f$ can be written as $a = b/f^n$ for some $n \geq 0$. Since the annihilator of f^{n+r} is $\mathfrak{q}_r$, the morphism $\phi : (f^{n+r}) \to I$ which maps f^{n+r}

to $f^r b$ is well defined (here (f^{n+r}) is the ideal of R generated by f^{n+r}). As I is injective, there is a map $\bar{\phi} : R \to I$ which extends ϕ:

$$
\begin{array}{ccc}
(f^{n+r}) & \lhook\joinrel\longrightarrow & R \\
{\scriptstyle \phi}\downarrow & \;\;\nearrow {\scriptstyle \bar{\phi}} & \\
I & &
\end{array}
$$

If $c = \bar{\phi}(1)$ then $f^{n+r}c = f^r b$. This means that c is a counterimage of $a = b/f^n$ in I. $\qquad\square$

Proof of Proposition 4.64. While we shall follow Hartshorne, the notes [50] may be very useful. This proof will use noetherian induction, i.e., the following principle [10, Lemma 1, p. 98], which we state in a form suitable to our case:

if X is a noetherian topological space, let P be a property of closed subsets of X. Let us assume that for all closed subsets Y of X, if every closed subset of Y enjoys P, then Y enjoys P as well. Then P holds for X.

The property we shall consider is the following:

if Y is a closed subset of X, and I is an injective R-module such that $\text{Supp}(\tilde{I}) \subset Y$,[q] *then $\tilde{I}$ is flabby.*

Proposition 4.64 is indeed this property for the space X. So, by the noetherian induction principle, we need to prove that for any closed subset $Y \subset X$, if the property is true for every proper closed subset of Y, then it is true for Y.

However, note that we can assume that Y is the closure of $\text{Supp}(\tilde{I})$, since otherwise $\text{Supp}(\tilde{I})$ is contained in a proper closed subset of Y, and then $\tilde{I}$ is flabby by the inductive hypothesis. If Y is a closed point, then $\tilde{I}$ is a skyscraper sheaf, which is flabby (Example 4.14), so that we are done. In general, we need to show that for every open subset U in X, the restriction morphism $\Gamma(X, \tilde{I}) \to \Gamma(U, \tilde{I})$ is surjective. Now we intersect with Y; if the intersection is empty there is nothing to prove, otherwise one can find $f \in R$ such that the open set $D(f)$ (see Example 4.51) is contained in

---

[q]See Definition 4.36.

U and intersects Y. We consider the diagram of morphisms

$$\Gamma_Z(X, \tilde{I}) \xrightarrow{\alpha} \Gamma_Z(U, \tilde{I}) \qquad (4.19)$$

$$\Gamma(X, \tilde{I}) \longrightarrow \Gamma(U, \tilde{I}) \longrightarrow \Gamma(D(f), \tilde{I})$$

where $Z = X - D(f)$, and Γ_Z denotes the submodule of sections with support contained in Z. Moreover, α is the natural restriction morphism. Note that $\Gamma(X, \tilde{I}) = I$ and $\Gamma(D(f), \tilde{I}) = I_f$.

Let $s \in \Gamma(U, \tilde{I})$. By Lemma 4.67 the morphism $\Gamma(X, \tilde{I}) \to \Gamma(D(f), \tilde{I})$ is surjective; take $t \in \Gamma(X, \tilde{I})$ that maps to $s_{|D(f)}$. Then $s - t_{|U}$ is zero in $D(f)$, i.e.,

$$\mathrm{Supp}(s - t_{|U}) \subset Z.$$

We show below that the morphism α in diagram (4.19) is surjective. Then, if $\alpha(u) = s - t_{|U}$, let $v = u + t$, so that

$$v_{|U} = s - t_{|U} + t_{|U} = s,$$

and the claim is proved.

We show that α is surjective. By Lemma 4.66, the R-module $J = \Gamma_Z(X, \tilde{I}) = \Gamma_{(f)}(I)$ is injective; moreover $\mathrm{Supp}(\tilde{J}) \subset Y \cap Z$. Now, $Y \cap Z$ is strictly contained in Y, so that, by the hypothesis of noetherian induction, $\tilde{J}$ is flabby. Thus $\Gamma(X, \tilde{J}) \to \Gamma(U, \tilde{J})$ is surjective; but on the other hand $\Gamma(U, \tilde{J}) = \Gamma_Z(U, \tilde{I})$ and $\Gamma(X, \tilde{J}) = \Gamma_Z(X, \tilde{I})$ (see Exercise II.5.6 in [28] for some details). $\qquad \square$

Now we have the tools to prove the main result of this section, namely, that the Čech cohomology of an open affine cover $\mathfrak{U}$ with coefficients in a quasi-coherent sheaf $\mathcal{F}$ on a (noetherian separated) scheme X is isomorphic to the sheaf cohomology of $\mathcal{F}$. We shall first prove that the sheaf cohomology vanishes in positive degree when X is affine; then, heuristically, one can expect that the sheaf cohomology of a quasi-coherent sheaf $\mathcal{F}$ on a nonaffine scheme depends on how the restrictions of $\mathcal{F}$ to the affine patches glue, and this, as we know, is described by Čech cohomology.

Theorem 4.68. *All quasi-coherent sheaves on the spectrum of a noetherian ring are acyclic.*

Proof. Given a quasi-coherent sheaf $\mathcal{F}$ on $X = \operatorname{Spec} R$, let $M = \Gamma(X, \mathcal{F})$. Let $I^\bullet$ be an injective resolution of M,

$$0 \to M \to I^\bullet;$$

by applying the exact functor $\sim$ we obtain a resolution $0 \to \mathcal{F} \to \tilde{I}^\bullet$ of $\mathcal{F}$, which is flabby by Lemma 4.64, and therefore, by Theorem 4.22, computes the cohomology of $\mathcal{F}$. So

$$H^i(X, \mathcal{F}) \simeq H^i(\Gamma(X, \tilde{I}^\bullet)) \simeq H^i(I^\bullet).$$

But $I^\bullet$ is exact in positive degree, so that the claim follows. $\qquad\square$

Actually this vanishing characterizes affine schemes.

Theorem 4.69 (Serre [56], see also [24, Theorem II.5.1.2] or [28, Theorem III.3.7]). *Let X be a noetherian scheme. The following conditions are equivalent:*

(1) *X is affine;*
(2) *$H^i(X, \mathcal{F}) = 0$ for all $i > 0$ and every quasi-coherent sheaf $\mathcal{F}$;*
(3) *$H^1(X, \mathcal{I}) = 0$ for all coherent sheaves of ideals $\mathcal{I}$ of the structure sheaf $\mathcal{O}_X$.*

Proof. The first condition implies the second by Theorem 4.68, and the third is a particular case of the second. So we only need to prove that the third implies the first. We do that in two steps.

Step 1. We prove that X can be covered by a finite number of open affine subsets of the form

$$X_f = \{p \in X \text{ such that } f_p \notin \mathfrak{m}_p\}$$

with $f \in \Gamma(X, \mathcal{O}_X)$, where $\mathfrak{m}_p$ is the maximal ideal of the local ring $\mathcal{O}_{X,p}$ (see Example 4.50). Take $p \in X$ a closed point, U an open affine neighbourhood of p and $Y = X - U$. The following diagram is commutative:

$$
\begin{array}{ccccccccc}
0 & \longrightarrow & \mathcal{I}_p & \longrightarrow & \mathcal{O}_X & \longrightarrow & i_*\mathcal{O}_p & \longrightarrow & 0 \\
 & & \uparrow & & \uparrow & & \| & & \\
0 & \longrightarrow & \mathcal{I}_{Y\cup\{p\}} & \longrightarrow & \mathcal{I}_Y & \longrightarrow & k(p) & \longrightarrow & 0 \\
 & & \uparrow & & \uparrow & & \| & & \\
 & & 0 & & 0 & & \mathcal{O}_{X,p}/\mathfrak{m}_p & &
\end{array}
$$

with horizontal exact sequences of $\mathcal{O}_X$-modules, where $k(p)$ is the skyscraper sheaf whose stalk at p is the residue field at p, and $\mathcal{I}_{Y\cup\{p\}}$, $\mathcal{I}_Y$ and $\mathcal{I}_p$ are the ideal sheaves of the closed subschemes $Y\cup\{p\}$, Y and p, respectively. The long exact sequence of cohomology of the second exact sequence contains the segment

$$\Gamma(X, \mathcal{I}_Y) \to \Gamma(X, k(p)) \to H^1(X, \mathcal{I}_{Y\cup\{p\}}) = 0,$$

where the equality holds by hypothesis, since $\mathcal{I}_{Y\cup\{p\}}$ is coherent as X is noetherian, see [28, Proposition II.5.9].

Take an $f \in \Gamma(X, \mathcal{I}_Y)$ which goes to $1 \in \Gamma(X, k(p)) = k(p)$. Note that $f \in \Gamma(X, \mathcal{O}_X)$ since there exists an inclusion $\Gamma(X, \mathcal{I}_Y) \hookrightarrow \Gamma(X, \mathcal{O}_X)$, moreover $p \in X_f$, and X_f is affine since $X_f = U_{\bar{f}} = D_{\bar{f}}$ where $\bar{f} = f_{|_U}$ and $f_{|_Y} = 0$. Since X is compact[r] (being noetherian) we can extract a finite number of such X_f. Let them be $X_{f_1}, \ldots, X_{f_r}$.

Step 2. We check that $f_1, \ldots, f_r$ generate R. By [28, Exercise II. 2.17(b)][s] this implies that X is affine. Define $\alpha : \mathcal{O}_X^{\oplus r} \to \mathcal{O}_X$ by $\alpha(a_1, \ldots, a_r) = \sum a_i f_i$. The morphism α is surjective since the X_{f_i} cover X. Let $\mathcal{G}$ be the kernel of α,

$$0 \to \mathcal{G} \to \mathcal{O}_X^{\oplus r} \xrightarrow{\alpha} \mathcal{O}_X \to 0.$$

The filtration

$$\mathcal{O}_X^{\oplus r} \supset \mathcal{O}_X^{\oplus(r-1)} \supset \cdots \supset \mathcal{O}_X \supset 0$$

induces a filtration

$$\mathcal{G} = \mathcal{G} \cap \mathcal{O}_X^{\oplus r} \supset \mathcal{G} \cap \mathcal{O}_X^{\oplus(r-1)} \supset \cdots \supset \mathcal{G} \cap \mathcal{O}_X \supset 0$$

whose successive quotients $\mathcal{G}_k$ are coherent[t] $\mathcal{O}_X$-submodules of

$$\mathcal{O}_X^{\oplus(k+1)}/\mathcal{O}_X^{\oplus k} = \mathcal{O}_X,$$

i.e., coherent sheaves of ideals of $\mathcal{O}_X$.[u]

---

[r]By compactness here we mean the usual property that any open cover has a finite subcover. Often in algebraic geometry this property is called *quasi-compactness*.
[s]A scheme X is affine if and only if there are $f_1, \ldots, f_r \in R = \Gamma(X, \mathcal{O}_X)$ that generate R, and the X_{f_i} are affine.
[t]This holds true because of [28, Proposition II.5.7].
[u]Note that the quotients $\mathcal{G}_k$ are coherent, and not only quasi-coherent, as they locally correspond to submodules of a finitely generated module over a noetherian ring (see for instance [41, Proposition 1.1]).

This implies that $H^1(X, \mathcal{G}) = 0$. But then the morphism

$$\alpha : \Gamma(X, \mathcal{O}_X^{\oplus r}) \to \Gamma(X, \mathcal{O}_X)$$

is surjective, and $f_1, \ldots, f_r$ generate the unit ideal in R. $\square$

The following result is a straightforward application of the Leray Theorem 4.33, and shows the relevance of Čech cohomology in scheme theory: sheaf cohomology may be replaced by the Čech cohomology of an affine cover, which is in principle easier to compute explicitly.

Theorem 4.70. *Let X be a noetherian separated scheme and $\mathfrak{U}$ an open affine cover of X. If $\mathcal{F}$ is a quasi-coherent sheaf on X, the morphisms (4.10) are isomorphisms in all degrees.*

Proof. Since X is separated, by Exercise II.4.3 in [28] all nonvoid intersections $U_{i_0 \ldots i_p}$ are affine. By Theorem 4.69 we have $H^k(U_{i_0 \ldots i_p}, \mathcal{F}) = 0$ for all p and all $k \geq 1$. Then the claim follows from the Leray Theorem 4.33.
$\square$

Hartshorne gives a slightly different proof (see [28, Theorem III.4.5]), which does not appeal directly to the Leray Theorem. This requires to embed $\mathcal{F}$ into a quasi-coherent flabby sheaf. This is possible due to the following result, which we include here for completeness.

Proposition 4.71. *A quasi-coherent sheaf $\mathcal{F}$ on a noetherian scheme X can be embedded into a quasi-coherent flabby sheaf.*

We already know that $\mathcal{F}$ can be embedded into a flabby sheaf, but the point is that one wants the flabby sheaf to be quasi-coherent.

Proof. Pick up a finite affine open cover $\{U_i = \operatorname{Spec} R_i\}$, and let $\mathcal{F}|_{U_i} \simeq \tilde{M}_i$. Choose embeddings into injective modules

$$0 \to M_i \to I_i \qquad\qquad (4.20)$$

and, denoting by $f_i : U_i \to X$ the inclusions, let

$$\mathcal{G} = \bigoplus_i f_{i*}(\tilde{I}_i).$$

By applying the localization functor $\sim$ to (4.20) one obtains the embeddings $\phi_i : \mathcal{F}|_{U_i} \to \tilde{I}_i$. Note also that for every open $U \subset X$

$$\mathcal{G}(U) = \bigoplus_i \tilde{I}_i(U \cap U_i),$$

which implies that $\mathcal{G}$ is quasi-coherent. Thus we can define an embedding $\phi : \mathcal{F} \to \mathcal{G}$ by letting, for every open $U \subset X$ and every $s \in \mathcal{F}(U)$,

$$\phi_U(s) = \sum_i \phi_i(s_{|U \cap U_i}).$$

Since every $\tilde{I}_i$ is flabby by Lemma 4.64, $\mathcal{G}$ is flabby. $\qquad\qquad\square$

4.6. Additional Exercises

1. (a) Prove that a quotient of a divisible abelian group is divisible.
   (b) Prove that for every left exact functor $F : \mathfrak{Ab} \to \mathfrak{B}$, where $\mathfrak{Ab}$ is the category of abelian groups and $\mathfrak{B}$ is any abelian category, one has $R^i F = 0$ for $i > 1$.
2. Check that the long exact sequence of cohomology associated with the sequence (3.3) splits into

$$0 \to H^0(Y, G_Y) \to G \oplus G \to G^{\oplus N} \to H^1(Y, G_Y) \to 0$$
$$0 \to H^n(Z, G_Z) \to H^{n+1}(Y, G_Y) \to 0, \quad n \geq 1,$$

   where N is the number of connected components of Z. If Y is connected and locally connected, and $N = 2$, and G is free over $\mathbb{Z}$, the first equation implies that $H^1(Y, G_Y) \simeq G$, thus showing by way of example that a constant sheaf on a reducible space need not be flabby. In this example, the reason is that Y is connected but contains two disjoint open subsets.
3. Prove that every separated presheaf of abelian groups embeds into an injective presheaf (see Exercise 17 in Chapter 3).
4. Let X be a topological space, U an open subset, $\mathcal{F}$ a sheaf of abelian groups on X, and $\mathcal{G}^\bullet$ a flabby resolution of $\mathcal{F}$.

   (a) Prove that $\mathcal{G}^\bullet_{|U}$ is a flabby resolution of $\mathcal{F}_{|U}$.
   (b) Show that this induces morphisms $H^\bullet(X, \mathcal{F}) \to H^\bullet(U, \mathcal{F}_{|U})$.[v]

---

[v] If $\mathfrak{U}$ is an open cover of X, we may use these restriction morphisms to define, for every $q \geq 0$, a complex of groups whose pth terms is

$$CH_q^p(\mathfrak{U}, \mathcal{F}) = \prod_{i_0 < \cdots < i_p} H^q(U_{i_0 \cdots i_p}, \mathcal{F}_{|U_{i_0 \cdots i_p}}),$$

with a Čech-type differential induced by the restriction morphisms mentioned above. The resulting cohomology complex is the first page of the spectral sequence described in Section 5.6.3.

5. Let X be a topological space, Z a closed subset, and $\mathcal{F}$ a sheaf of abelian groups on X. Define a "restriction" morphism $H^\bullet(X, \mathcal{F}) \to H^\bullet(Z, \mathcal{F}_{|Z})$.

6. Spell out the details of the induction step in the proof of the Leray Theorem 4.33.

7. (De Rham cohomology of the circle) Let t be a coordinate on the circle S^1 varying between 0 and 1.

   (a) Show that a differential 1-form $\omega = f(t)\, dt$ on S^1 is exact if and only if $\int_0^1 f(t)\, dt = 0$.

   (b) Use this fact to compute $H^1_{dR}(S^1)$.

8. The nth Hirzebruch complex surface[w] $\mathbb{F}_n$ is a locally trivial $\mathbb{P}^1$ fibration over $\mathbb{P}^1$ (where $\mathbb{P}^1$ is the complex projective line, see Example 4.62). If (x, y) are homogeneous coordinates on $\mathbb{P}^1$, let $U = \{x \neq 0\}$, $V = \{y \neq 0\}$, and $\mathfrak{U} = (U, V)$. Denote $\pi : \mathbb{F}_n \to \mathbb{P}^1$ the projection, so that $\pi^{-1}(U) \simeq \pi^{-1}(V) \simeq \mathbb{C} \times \mathbb{P}^1$, $\pi^{-1}(U \cap V) \simeq \mathbb{C}^* \times \mathbb{P}^1$. Use the Mayer–Vietoris sequence (see Example 4.39) to compute the cohomology groups

$$H^1(\mathbb{F}_n, \mathbb{R}) \simeq H^3(\mathbb{F}_n, \mathbb{R}) = 0,$$

$$H^2(\mathbb{F}_n, \mathbb{R}) \simeq \mathbb{R} \oplus \mathbb{R}, \quad H^4(\mathbb{F}_n, \mathbb{R}) \simeq \mathbb{R}.$$

Hint: you will need to use the Poincaré duality $H^3(\mathbb{F}_n, \mathbb{R}) \simeq H^1(\mathbb{F}_n, \mathbb{R})^*$.

9. (De Rham cohomology of the spheres) Use the Mayer–Vietoris sequence (see Example 4.39) to show that the de Rham cohomology of the sphere S^n, $n \geq 1$, is

$$H^0_{dR}(S^n) \simeq H^n_{dR}(S^n) \simeq \mathbb{R}, \quad H^k_{dR}(S^n) = 0 \quad \text{if } 0 < k < n.$$

Hint: use induction on n.

10. Let $(X, \mathcal{O}_X)$ be a ringed space, and $\mathcal{G}$ an $\mathcal{O}_X$-module. Prove that

$$\mathcal{H}om_{\mathcal{O}_X}(\mathcal{O}_X, \mathcal{G}) \simeq \mathcal{G}.$$

---

[w] See [2, Section V.4].

11. Let $(X, \mathcal{O}_X)$ be a ringed space, and let $\mathcal{F}$, $\mathcal{G}$ be $\mathcal{O}_X$-modules.

(a) Prove that when $\mathcal{F}$ is locally free of finite rank,[x] then

$$\mathcal{H}om_{\mathcal{O}_X}(\mathcal{F}, \mathcal{G}) \simeq \mathcal{F}^\vee \otimes_{\mathcal{O}_X} \mathcal{G},$$

where $\mathcal{F}^\vee = \mathcal{H}om_{\mathcal{O}_X}(\mathcal{F}, \mathcal{O}_X)$ is the dual $\mathcal{O}_X$-module of $\mathcal{F}$.

(b) Prove that if $\mathcal{M}$ is a locally free $\mathcal{O}_X$-module of finite rank, then

$$\mathcal{E}xt^i_{\mathcal{O}_X}(\mathcal{F} \otimes_{\mathcal{O}_X} \mathcal{M}, \mathcal{G}) \simeq \mathcal{E}xt^i_{\mathcal{O}_X}(\mathcal{F}, \mathcal{M}^\vee \otimes_{\mathcal{O}_X} \mathcal{G})$$

$$\simeq \mathcal{E}xt^i_{\mathcal{O}_X}(\mathcal{F}, \mathcal{G}) \otimes_{\mathcal{O}_X} \mathcal{M}^\vee \quad \text{for } i \geq 0.$$

Hint: you will need to use Exercise 11.

12. Let $(X, \mathcal{O}_X)$ be a ringed space. Prove that if $\mathcal{I}$ is an injective $\mathcal{O}_X$-module, and $\mathcal{M}$ is a locally free $\mathcal{O}_X$-module of finite rank, then $\mathcal{M} \otimes_{\mathcal{O}_X} \mathcal{I}$ is injective.

Hint: the functor $\mathcal{M}^\vee \otimes_{\mathcal{O}_X} -$ is exact.

13. $\mathbb{P}^1$ is the one-dimensional complex projective space (see Example 4.62). Let U and V be the usual open subsets on $\mathbb{P}^1$ associated with a choice of homogeneous coordinates. Prove that $H^1(\mathfrak{U}, \mathcal{O}) = 0$, where $\mathcal{O}$ is the sheaf of holomorphic functions on $\mathbb{P}^1$. Deduce that $H^1(\mathbb{P}^1, \mathcal{O}) = 0$.

Hint: using affine coordinates $z = y/x$ and $t = x/y$ on U and V, respectively, write Laurent expansions for the holomorphic functions on U, V and $U \cap V$.

14. Let $\mathbb{P}^1_\Bbbk$ be the one-dimensional projective space over an algebraically closed field $\Bbbk$ (see p. 110). Denote by $\mathcal{O}$ the sheaf of regular functions on it, and for a given closed point $p \in \mathbb{P}^1_\Bbbk$, let $\mathcal{O}(-p)$ be the sheaf of regular functions that vanish at p, i.e., the kernel of the evaluation morphism $\mathcal{O} \to \Bbbk_p$, where $\Bbbk_p = \mathcal{O}_p/\mathfrak{m}_p$ is a skyscraper sheaf supported at p.[y] There is an exact sequence

$$0 \to \mathcal{O}(-p) \to \mathcal{O} \to \Bbbk_p \to 0$$

of which we consider the long exact sequence of cohomology.

---

[x] An $\mathcal{O}_X$-module $\mathcal{M}$ is locally free of rank r if it is locally isomorphic to $\mathcal{O}_X^{\oplus r}$.

[y] The sheaves $\mathcal{O}(-p)$, with varying p, are all isomorphic, and are usually denoted $\mathcal{O}(-1)$. The dual line bundle is denoted $\mathcal{O}(1)$.

(a) Prove that the morphism $H^0(\mathbb{P}^1_{\Bbbk}, \mathcal{O}) \to H^0(\mathbb{P}^1_{\Bbbk}, \Bbbk_p)$ is an isomorphism.

(b) Compute the cohomology groups $H^i(\mathbb{P}^1_{\Bbbk}, \mathcal{O}(-p))$.

(c) For every $n \geq 1$, let $\mathcal{O}(-np)$ be the sheaf of regular functions that have a zero of order n at p. Compute the groups $H^0(\mathbb{P}^1_{\Bbbk}, \mathcal{O}(-np))$.

Hint: the cohomology of $\mathcal{O}$ is $H^0(\mathbb{P}^1_{\Bbbk}, \mathcal{O}) = \Bbbk$, $H^i(\mathbb{P}^1_{\Bbbk}, \mathcal{O}) = 0$ for $i > 0$ (see [28, Section III.5]).

15. Denote by $\mathbb{P}^n$ the n-dimensional projective space over the complex numbers. The Euler exact sequence on $\mathbb{P}^n$ is

$$0 \to \mathcal{O} \to \mathcal{O}(1)^{\oplus(n+1)} \to T_{\mathbb{P}^n} \to 0,$$

where $T_{\mathbb{P}^n}$ is the tangent bundle to $\mathbb{P}^n$.

(a) Use the previous exact sequence to compute the sheaf cohomology groups of $T_{\mathbb{P}^n}$.

(b) After dualizing the Euler sequence, compute the sheaf cohomology groups of the cotangent bundle $\Omega^1_{\mathbb{P}^n}$.

(c) Using the Hodge Decomposition Theorem (see for instance [62, Proposition 6.11])

$$H^m(\mathbb{P}^n, \mathbb{C}) \simeq \bigoplus_{p+q=m} H^p(\mathbb{P}^n, \Omega^q_{\mathbb{P}^n}),$$

compute the groups $H^m(\mathbb{P}^n, \mathbb{C})$.

16. Let $\mathcal{O}^*$ be the sheaf of nowhere vanishing holomorphic functions on a complex projective space $\mathbb{P}^n$. Prove that $H^1(\mathbb{P}^n, \mathcal{O}^*) \simeq \mathbb{Z}$.
    Hint: you will need Proposition 4.34; for the singular cohomology of $\mathbb{P}^n$ see, e.g., [29, Theorem 3.12].

17. Let $\mathfrak{p}$ be a prime ideal in a commutative ring with unity R. Prove that the set $V(\mathfrak{p})$ is the closure of the point $\mathfrak{p}$ in $\operatorname{Spec} R$.

18. Prove that if R is a noetherian ring every subset of $\operatorname{Spec} R$ is compact.[z]
    Hints: (1) it is enough to consider open subsets;
    (2) every ideal of R is finitely generated.

19. (a) Prove that the kernel of a morphism of quasi-coherent sheaves on a scheme is quasi-coherent.

---

[z] See footnote on p. 117.

(b) Prove that on a noetherian scheme the kernel of a morphism of coherent sheaves is coherent.

(c) Let $\phi : R \to S$ be a morphism of commutative rings with unity, where R is noetherian, let $X = \operatorname{Spec} R$ and $Y = \operatorname{Spec} S$, and let $f : Y \to X$ be the induced scheme morphism. Assume that the morphism $f^{\sharp} : \mathcal{O}_X \to f_* \mathcal{O}_Y$ is surjective. Prove that ϕ is surjective.

20. For each of the following rings, say if the associated spectrum is irreducible and/or reduced (see Remark 4.59):

- $\Bbbk[x, y]/(xy)$;
- $\Bbbk[x, y]/(y^2)$;
- $\Bbbk[x, y]/(x^2, y)$;
- $\Bbbk[x, y]/(x - y^2)$;
- $\Bbbk[x, y]/(y^2 - x^3 - x^2)$.

21. Prove that the scheme $X = \mathbb{A}_{\Bbbk}^2 \setminus \{0\}$ is not affine by showing that $H^1(X, \mathcal{O}_X) \neq 0$.

Hint: if $\mathbb{A}_{\Bbbk}^2 = \operatorname{Spec} \Bbbk[x, y]$, then $\mathfrak{U} = (U, V)$, with $U = \{x \neq 0\}$, $V = \{y \neq 0\}$ (i.e., $U = D(x)$, $V = D(y)$), is an affine open cover of X, so that by Theorem 4.70 one can compute $H^1(\mathfrak{U}, \mathcal{O}_X)$. Do that explicitly.[aa]

22. Let X be a differentiable manifold, a complex manifold, or a scheme, and let $\mathcal{O}_X$ be the corresponding sheaf of functions. A line bundle $\mathcal{L}$ on X is a locally free $\mathcal{O}_X$-module of rank 1, i.e., $\mathcal{L}$ is locally isomorphic to $\mathcal{O}_X$. We say that $\mathcal{L}$ trivializes on an open cover $\mathfrak{U} = \{U_i\}$ if $\mathcal{L}(U_i) \simeq \mathcal{O}_X(U_i)$. This defines, for any nonvoid intersection $U_i \cap U_j$, sections $g_{ij} \in \mathcal{O}_X^*(U_i \cap U_j)$, called *transition functions*.

(a) Show that the transition functions define a cocycle in $C^1(\mathfrak{U}, \mathcal{O}_X^*)$ (therefore, they determine a class in $H^1(\mathfrak{U}, \mathcal{O}_X^*)$).

(b) Prove that two line bundles which trivialize on the same open cover $\mathfrak{U}$ are isomorphic if and only if they determine the same class in $H^1(\mathfrak{U}, \mathcal{O}_X^*)$.

[aa] Another argument which does not use cohomology is the following. By Hartog's lemma $\Gamma(X, \mathcal{O}_X) = \Gamma(\mathbb{A}_{\Bbbk}^2, \mathcal{O}_{\mathbb{A}^2}) = \Bbbk[x, y]$, so that if X were affine, it would be isomorphic to $\mathbb{A}_{\Bbbk}^2$, which of course is not true. The holomorphic Hartog's Lemma states that any holomorphic function on the punctured open n-polydisc, $n \geq 2$, extends to the whole polydisc, see, e.g., [21]; for the algebraic version see [49, Remark 31.16.2].

(c) Assume that X is a differentiable or complex manifold, and let $\nu : H^1(\mathfrak{U}, \mathcal{O}_X^*) \to H^2(\mathfrak{U}, \mathbb{Z})$ be the second connecting morphism of the exponential sheaf sequence; the *first Chern class* $c_1(\mathcal{L})$ is obtained by applying ν to the class of $\mathcal{L}$ in $H^1(\mathfrak{U}, \mathcal{O}_X^*)$. Show that $c_1(\mathcal{L})$ is represented by the cocycle

$$\frac{1}{2\pi i}(\log g_{ij} + \log g_{jk} + \log g_{ki}).$$

Chapter 5

Spectral Sequences

Spectral sequences are a powerful tool for computing homology, cohomology and homotopy groups. Often they allow one to trade a difficult computation for an easier one. Examples that we shall consider are another proof of the Čech–de Rham Theorem, the spectral sequence associated to an open cover (Čech spectral sequence), the Leray spectral sequence, the local-to-global spectral sequence, the Frölicher spectral sequence of a complex manifold, the Hochschild–Serre spectral sequence associated with an extension of Lie algebras, and the Künneth spectral sequences. Moreover, we shall introduce hypercohomology, and, more generally, hyperderived functors, with the associated spectral sequences. Many of these spectral sequences will be special cases of *Grothendieck's spectral sequence*, which relates the derived functors of a composition of functors with the composition of the derived functors.

Spectral sequences are a difficult topic, basically because the theory is quite intricate and the notation is correspondingly cumbersome. Therefore, we have chosen what seems to us to be the simplest approach, due to Massey [44]. Initially our treatment basically follows [8].

For simplicity in Sections 5.1–5.3, we deal with spectral sequences in the category of modules, but the same treatment applies verbatim in any abelian category admitting infinite direct sums.

5.1. Filtered Complexes

The fundamental object giving rise to a spectral sequence is a *filtered complex*, i.e., a differential complex equipped with a filtration which is compatible with the grading and with the differential of the complex, in the sense we explain below.

Let R be a commutative ring, and let $(K^{\bullet}, d)$ be a complex in R-**mod**, so that

$$K = \bigoplus_{n \in \mathbb{Z}} K^n, \quad d \colon K^n \to K^{n+1}, \quad d^2 = 0.$$

A subcomplex of $(K^{\bullet}, d)$ is a graded submodule $K' \subset K$ such that $dK' \subset K'$, that is,

$$K' = \bigoplus_{n \in \mathbb{Z}} K'^n, \quad K'^n \subset K^n, \quad d \colon K'^n \to K'^{n+1}.$$

A sequence of nested subcomplexes

$$K = K_0 \supset K_1 \supset K_2 \supset \cdots$$

is a *filtration* of $(K^{\bullet}, d)$. We then say that $(K^{\bullet}, d)$ is filtered, and associate with it the *graded complex*[a]

$$\mathrm{Gr}(K) = \bigoplus_{p \in \mathbb{Z}} K_p/K_{p+1}, \quad K_p = K \text{ if } p \leq 0.$$

Note that by assumption (since every K_{p+1} is a graded submodule of K_p) the filtration is compatible with the grading, i.e., if we define $K_p^n = K^n \cap K_p$, then

$$K^n = K_0^n \supset K_1^n \supset K_2^n \supset \cdots \tag{5.1}$$

is a filtration of K^n, and moreover $dK_p^n \subset K_p^{n+1}$.

Example 5.1. A (first quadrant) *double complex* is a collection of R-modules $K^{p,q}$, with $p, q \geq 0$,[b] and morphisms $\delta_1 \colon K^{p,q} \to K^{p+1,q}$, $\delta_2 \colon K^{p,q} \to K^{p,q+1}$ such that

$$\delta_1{}^2 = \delta_2{}^2 = 0, \quad \delta_1 \delta_2 + \delta_2 \delta_1 = 0.$$

Let $(T^{\bullet}, d)$ be the associated *total complex* (cf. footnote e in Chapter 2):

$$T^n = \bigoplus_{p+q=n} K^{p,q}, \quad d \colon T^n \to T^{n+1} \text{ defined by } d = \delta_1 + \delta_2$$

---

[a]The choice of having $K_p = K$ for $p \leq 0$ is due to notational convenience.
[b]This assumption is made here for simplicity but one could let p, q range over the integers; however, some of the results we are going to give would be no longer valid.

(note that the definition of d implies $d^2 = 0$). Then, letting

$$T_p = \bigoplus_{n \geq p, q \geq 0} K^{n,q},$$

we obtain a filtration of $(T^\bullet, d)$. This satisfies $T_p \simeq T$ for $p \leq 0$. The successive quotients of the filtration are

$$\mathrm{Gr}(T)_p = T_p/T_{p+1} = \bigoplus_{q \in \mathbb{N}} K^{p,q}.$$

This filtration is called *the filtration by columns*. Figure 5.1 depicts this filtration.

Analogously, the *filtration by rows* is

$$T'_q = \bigoplus_{p \geq 0, \, n \geq q} K^{p,n},$$

and the corresponding graded module is

$$\mathrm{Gr}(T')_q = T'_q/T'_{q+1} = \bigoplus_{p \in \mathbb{N}} K^{p,q}.$$

Definition 5.2. A filtration $K_\bullet$ of $(K^\bullet, d)$ is regular if for every $n \geq 0$ the filtration (5.1) is finite; in other words, for every n there is a number $\ell(n)$ such that $K_p^n = 0$ for $p > \ell(n)$.

For instance, the filtration by columns in Example 5.1 is regular since $T_p^n = 0$ for $p > n$, as

$$T_p^n = T^n \cap T_p = \bigoplus_{j=0}^{n-p} K^{n-j,j}.$$

In the same way, the filtration by rows is regular as well as

$$T_q^n = T^n \cap T_q = \bigoplus_{j=0}^{n-q} K^{j,n-j}.$$

5.2. The Spectral Sequence of a Filtered Complex

Now we shall see how a filtered complex K gives rise to a spectral sequence, i.e., a sequence of differential complexes $\{E_r\}$ such that the cohomology of E_r is isomorphic to the complex E_{r+1}. Under appropriate hypotheses, this procedure stabilizes, and one says that the spectral sequence *converges*, actually to the cohomology of the complex K, or more precisely, to the graded module of a suitable filtration of it.

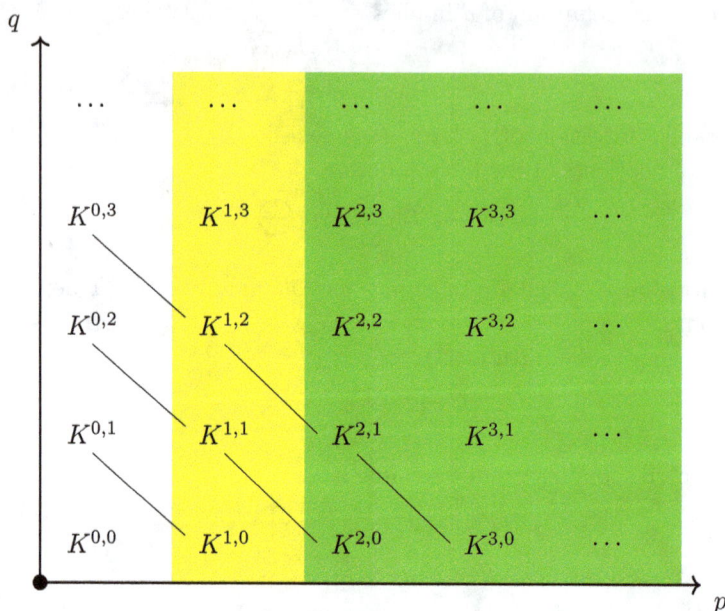

Figure 5.1. A first quadrant double complex. The homogeneous summands of the total complex are the sums of the terms connected by the skew lines. The T_2 term of the filtration by columns is the sum of the terms in the green area, extending indefinitely to the right and vertically. The T_1 term is the sum of the terms in the yellow and green areas. The yellow area represents the term $\mathrm{Gr}(T)_1$ of the graded module. The filtrations of the homogeneous summands are obtained by discarding the successive terms from the sums along the skew lines starting from the lower right end.

5.2.1. The spectral sequence

Here we construct the spectral sequence. At first we shall not consider the grading. Let $K_\bullet$ be a compatible filtration of a differential module (K, d), i.e., $dK_p \subset K_p$, and let

$$G = \bigoplus_{p \in \mathbb{Z}} K_p.$$

The inclusions $K_{p+1} \to K_p$ induce a morphism $i \colon G \to G$ ("the shift by the filtering degree"), and one has an exact sequence

$$0 \to G \xrightarrow{i} G \xrightarrow{j} E \to 0 \tag{5.2}$$

with $E \simeq \mathrm{Gr}(K)$. The differential d induces differentials in G and E, so that from (5.2) one gets an exact triangle in cohomology

(see Proposition 1.20)

$$H(G) \xrightarrow{\quad i \quad} H(G) \qquad (5.3)$$

$$\downarrow k \qquad \nearrow j$$

$$H(E)$$

where k is the connecting morphism.

Let us now assume that the filtration $K_\bullet$ has *finite length*, i.e., $K_p = 0$ for p greater than some ℓ (called the *length* of the filtration). Since $dK_p \subset K_p$ for every p, we may consider the cohomology groups $H(K_p)$. The morphism i induces morphisms $i \colon H(K_{p+1}) \to H(K_p)$. Define G_1 to be the direct sum of the terms of the sequence (which is not exact)

$$0 \to H(K_\ell) \xrightarrow{i} H(K_{\ell-1}) \xrightarrow{i} \cdots \xrightarrow{i} H(K_1) \xrightarrow{i} H(K) \xrightarrow{\sim} H(K_{-1}) \xrightarrow{\sim} \cdots,$$

i.e., $G_1 = \bigoplus_{p\in\mathbb{Z}} H(K_p) \simeq H(G)$. Next we define G_2 as the sum of the terms of the sequence

$$0 \to i(H(K_\ell)) \to i(H(K_{\ell-1}))$$

$$\to \cdots \to i(H(K_1)) \to H(K) \xrightarrow{\sim} H(K_{-1}) \xrightarrow{\sim} \cdots.$$

Note that the morphism $i(H(K_1)) \to H(K)$ is injective, since it is the inclusion of the image of $i \colon H(K_1) \to H(K)$ into $H(K)$. This procedure is then iterated: G_3 is the sum of the terms in the sequence

$$0 \to i(i(H(K_\ell))) \to i(i(H(K_{\ell-1})))$$

$$\to \cdots \to i(i(H(K_2))) \to i(H(K_1)) \to H(K) \xrightarrow{\sim} H(K_{-1}) \xrightarrow{\sim} \cdots$$

and now the morphisms $i(i(H(K_2))) \to i(H(K_1))$ and $i(H(K_1)) \to H(K)$ are injective. When we reach the step $\ell + 1$, all the morphisms in the sequence

$$0 \to i^\ell(H(K_\ell)) \to i^{\ell-1}(H(K_{\ell-1}))$$

$$\to \cdots \to i(H(K_1)) \to H(K) \xrightarrow{\sim} H(K_{-1}) \xrightarrow{\sim} \cdots$$

are injective. Now $G_{\ell+2} \simeq G_{\ell+1}$, and the procedure stabilizes: $G_r \simeq G_{r+1}$ for $r \geq \ell + 1$. We define $G_\infty = G_{\ell+1}$; we have

$$G_\infty \simeq \bigoplus_{p\in\mathbb{Z}} F_p,$$

where $F_p = i^p(H(K_p))$, i.e., F_p is an injective image of $H(K_p)$ into $H(K)$. The groups F_p provide a filtration of $H(K)$,

$$H(K) = F_0 \supset F_1 \supset \cdots \supset F_\ell \supset F_{\ell+1} = 0. \qquad (5.4)$$

We come now to the construction of the spectral sequence. Recall that since $dK_p \subset K_p$, and $E = \bigoplus_p K_p/K_{p+1}$, the differential d acts on E, and one has a cohomology group $H(E)$ which splits into a direct sum

$$H(E) \simeq \bigoplus_{p \in \mathbb{Z}} H(K_p/K_{p+1}, d).$$

The cohomology group $H(E)$ fits into the exact triangle (5.3), that we rewrite as

$$
\begin{array}{ccc}
G_1 & \xrightarrow{\;i_1\;} & G_1 \\
& \searrow{\scriptstyle k_1} \quad \nearrow{\scriptstyle j_1} & \\
& E_1 &
\end{array}
\tag{5.5}
$$

where $E_1 = H(E)$. We define $d_1 \colon E_1 \to E_1$ by letting $d_1 = j_1 \circ k_1$; then $d_1^2 = 0$ since the triangle is exact. Let $E_2 = H(E_1, d_1)$ and recall that G_2 is the image of G_1 under $i \colon G_1 \to G_1$. We have morphisms

$$i_2 \colon G_2 \to G_2, \quad j_2 \colon G_2 \to E_2, \quad k_2 \colon E_2 \to G_2,$$

where

   (i) i_2 is induced by i_1 by letting $i_2(i_1(x)) = i_1(i_1(x))$ for $x \in G_1$;
  (ii) j_2 is induced by j_1 by letting $j_2(i_1(x)) = [j_1(x)]$ for $x \in G_1$, where $[\ ]$ denotes taking the cohomology class in $E_2 = H(E_1, d_1)$;
 (iii) k_2 is induced by k_1 by letting $k_2([y]) = k_1(y)$.

Proposition 5.3. *The morphisms i_2, j_2 and k_2 are well defined, and the triangle*

$$
\begin{array}{ccc}
G_2 & \xrightarrow{\;i_2\;} & G_2 \\
& \searrow{\scriptstyle k_2} \quad \nearrow{\scriptstyle j_2} & \\
& E_2 &
\end{array}
\tag{5.6}
$$

is exact.

Proof. Let us first prove that j_2 is well defined, i.e., $j_1(x)$ is closed and its cohomology class does not depend on the choice of x. Note indeed that $d_1 j_1(x) = j_1 k_1 j_1(x) = 0$. Furthermore, let $i_1(x) = i_1(x')$. Then, $i_1(x - x') = 0$, and $x - x' = k_1(y)$ for some $y \in E_1$, and

$$j_1(x) - j_1(x') = j_1 k_1(y) = d_1(y),$$

so that $[j_1(x)] = [j_1(x')]$.

Analogously, one proves that i_2 and k_2 are well defined.

We check the exactness at the G_2 on the left. The conditions $\operatorname{im} k_2 \subset \ker i_2$ follows from the equalities

$$i_2 k_2([y]) = i_1^3 k_1(y) = 0,$$

while $\ker i_2 \subset \operatorname{im} k_2$ because, if $i_2 i_1(x) = 0$, then $i_1 i_1(x) = 0$, so that

$$i_1(x) = k_1(y) = k_2([y]);$$

note that y is a cycle as

$$d_1(y) = j_1 k_1(y) = j_1 i_1(x) = 0.$$

Exactness at the G_2 on the right and at E_2 is proved by similar computations (the case of E_2 is detailed in [8, p. 156]). □

We call (5.6) the *derived triangle* of (5.5). The procedure leading from (5.5) to the triangle (5.6) can be iterated, and we get a sequence of exact triangles

$$
\begin{array}{ccc}
G_r & \xrightarrow{\;i_r\;} & G_r \\
& \underset{k_r}{\nwarrow} \quad \swarrow{\scriptstyle j_r} & \\
& E_r &
\end{array}
$$

where each R-module E_r is the cohomology of the differential module (E_{r-1}, d_{r-1}), with $d_{r-1} = j_{r-1} \circ k_{r-1}$.

As we have already noticed, due to the assumption that the filtration $K_\bullet$ has finite length ℓ, the modules G_r stabilize when $r \geq \ell + 1$, and the morphisms $i_r \colon G_r \to G_r$ become injective. Thus, all morphisms $k_r \colon E_r \to G_r$ vanish in that range, which implies $d_r = 0$, so that the modules E_r stabilize as well: $E_{r+1} \simeq E_r$ for $r \geq \ell + 1$. We denote by $E_\infty = E_{\ell+1}$ the stable value.

Thus, the sequence

$$0 \to G_\infty \xrightarrow{\;i_\infty\;} G_\infty \xrightarrow{\;j_\infty\;} E_\infty \to 0$$

is exact, which implies that E_∞ is the associated graded module of the filtration (5.4) of $H(K)$:

$$E_\infty \simeq \bigoplus_{p \leq \ell} F_p / F_{p+1}.$$

We introduce a suitable terminology for this construction.

Definition 5.4. A sequence of differential modules $\{(E_i, d_i)\}$ such that

$$H(E_i, d_i) \simeq E_{i+1}$$

is said to be a spectral sequence. If the modules E_i eventually become stationary, we denote the stationary value by E_∞. If E_∞ is isomorphic to the associated graded module of some filtered module H, we say that the spectral sequence converges to H.

So what we have seen so far in this section is that, if (K, d) is a differential module with a filtration of finite length, one can build a spectral sequence which converges to $H(K)$.

Remark 5.5. It may happen in special cases that the modules E_i stabilize before getting the value $i = \ell + 1$. That happens if and only if there is an integer r_0 such that $d_i = 0$ for $i \geq r_0$. When this happens we say that the spectral sequence *degenerates* at step r_0.

Now we consider the presence of a grading.

Theorem 5.6. *Let $(K^\bullet, d)$ be a complex of R-modules, and $K_\bullet$ a regular filtration. There is a spectral sequence $\{(E_i, d_i)\}$, where each E_i is graded, which converges to the graded module $H^\bullet(K, d)$.*

Note that the filtration need not be of finite length: the length $\ell(n)$ of the filtration of K^n is finite for every n, but may increase with n.

Proof. For every n and p we have $d(K_p^n) \subset K_p^{n+1}$, therefore we have cohomology modules $H^n(K_p)$. As a consequence, the modules G_r are graded:

$$G_r^n = \bigoplus_{p \in \mathbb{Z}} i^{r-1}(H^n(K_p))$$

and the modules E_r are accordingly graded. We may construct the derived triangles as before, but now we should pay attention to the grading: the morphisms i and j have degree zero, but k has degree one (just check the definition: k is basically a connecting morphism).

Fix a natural number n, and let $r \geq \ell(n) + 1$; the morphisms

$$i_r \colon G_r^{n+1} \to G_r^{n+1}$$

are injective, and the morphisms

$$k_r \colon E_r^n \to G_r^{n+1}$$

are zero. These are the same statements as in the ungraded case. Therefore, as it happened in that case, the modules E_r^n become stationary for r big enough. Note that $G_\infty^n = \bigoplus_{p\in\mathbb{Z}} F_p^n$, where

$$F_p^n = i^{\ell(n)}(H^n(K_p)),$$

that the morphism i_∞ sends F_{p+1}^n injectively into F_p^n for every n, and there is an exact sequence

$$0 \to G_\infty^n \xrightarrow{i_\infty} G_\infty^n \to E_\infty^n \to 0.$$

This implies that E_∞ is the graded module associated with the filtered module $H^\bullet(K,d)$. $\qquad\square$

The last statement in the proof means that for each n, $F_\bullet^n$ is a filtration of $H^n(K,d)$, and

$$E_\infty^n \simeq \bigoplus_{p\in\mathbb{Z}} F_p^n/F_{p+1}^n.$$

5.2.2. The bidegree and the five-term sequence

The terms E_i of the spectral sequence are actually bigraded; this will allow us to show that its lower degree terms are related by an exact sequence, which is often very useful in the applications. Since the filtration and the degree of $K^\bullet$ are compatible, we have

$$K_p/K_{p+1} \simeq \bigoplus_{q\in\mathbb{Z}} K_p^q/K_{p+1}^q \simeq \bigoplus_{q\in\mathbb{Z}} K_p^{p+q}/K_{p+1}^{p+q}$$

and $E_0 = E = \mathrm{Gr}(K)$ is bigraded by

$$E_0 = \bigoplus_{p,q\in\mathbb{Z}} E_0^{p,q} \quad \text{with } E_0^{p,q} = K_p^{p+q}/K_{p+1}^{p+q}. \tag{5.7}$$

Note that the total complex associated with this bidegree yields the grading of E:

$$\bigoplus_{p+q=n} E_0^{p,q} = \bigoplus_{p+q=n} K_p^{p+q}/K_{p+1}^{p+q} = K_p^n/K_{p+1}^n = E^n.$$

Let us go to the next step. Since $d: K_p^{p+q} \to K_p^{p+q+1}$, i.e., $d: E_0^{p,q} \to E_0^{p,q+1}$, and $E_1 = H(E,d)$, if we set

$$E_1^{p,q} = H^q(E_0^{p,\bullet}, d) \simeq H^{p+q}(K_p/K_{p+1})$$

we have $E_1 \simeq \bigoplus_{p,q\in\mathbb{Z}} E_1^{p,q}$.

If we go one step further we can show that

$$d_1 \colon E_1^{p,q} \to E_1^{p+1,q}.$$

Indeed if $x \in E_1^{p,q} \simeq H^{p+q}(K_p/K_{p+1})$ we write x as $x = [y]$, where $y \in K_p^{p+q}/K_{p+1}^{p+q}$, so that $k_1(x) = i_1(k(y)) \in H^{p+q+1}(K_{p+1})$, and

$$d_1(x) = j_1(k_1(x)) = j_1(k(y)) \in H^{p+q+1}(K_{p+1}/K_{p+2}) \simeq E_1^{p+1,q}.$$

As a result, we have $E_2 \simeq \bigoplus_{p,q \in \mathbb{Z}} E_2^{p,q}$ with

$$E_2^{p,q} \simeq H^p(E_1^{\bullet,q}, d_1).$$

The same analysis shows that in general $E_i \simeq \bigoplus_{p,q \in \mathbb{Z}} E_i^{p,q}$ with

$$d_i \colon E_i^{p,q} \to E_i^{p+i,q-i+1}$$

and moreover we have

$$E_\infty^{p,q} \simeq F_p^{p+q}/F_{p+1}^{p+q} \tag{5.8}$$

(see for instance [20, Section I.4; 62, Section 8.3.2] or [64, Section 5.2]). The case of d_2 is spelled out in detail at p. 139; this can also give an idea of the general case.

The next two lemmas establish the existence of the morphisms that we shall use to introduce the so-called *five-term sequence*, and will anyway be useful in the following.

Lemma 5.7. *For every $i \geq 1$, there are canonical morphisms $H^q(K) \to E_i^{0,q}$.*

Proof. Since $K_0 = K$, we have $E_\infty^{0,q} \simeq F_0^q/F_1^q \simeq H^q(K)/F_1^q$, so that there is a surjective morphism $H^q(K) \to E_\infty^{0,q}$.

Note now that a nonzero class in $E_i^{0,q}$ cannot be a boundary, since then it should come from $E_i^{-i,q+i-1}$ which is zero because $K_p = K$ for $p \leq 0$. So cohomology classes are cycles. Since cohomology classes are elements in $E_{i+1}^{0,q}$, we have inclusions $E_{i+1}^{0,q} \subset E_i^{0,q}$, i.e., $E_{i+1}^{0,q}$ is the submodule of cycles in $E_i^{0,q}$. This yields an inclusion $E_\infty^{0,q} \subset E_i^{0,q}$ for all i.

Combining the two arguments we obtain morphisms $H^q(K) \to E_i^{0,q}$. $\square$

Lemma 5.8. *Assume that $K_p^n = 0$ if $p > n$ (so, in particular, the filtration is regular). Then, for every $i \geq 2$, there is a morphism $E_i^{p,0} \to H^p(K)$.*

Proof. The hypothesis of the lemma implies that $E_i^{p,q} = 0$ for $q < 0$ (indeed, $F_p^{p+q} = i^r(H^{p+q}(K_p))$ for r big enough, so that $F_p^{p+q} = 0$ if $q < 0$ since then $K_p^{p+q} = 0$). As a consequence, for $i \geq 2$ the differential

$d_i \colon E_i^{p,0} \to E_i^{p+i,1-i}$ vanishes, i.e., all elements in $E_i^{p,0}$ are cycles, and therefore determine cohomology classes in $E_{i+1}^{p,0}$. This means that we have a morphism $E_i^{p,0} \to E_{i+1}^{p,0}$, and composing, morphisms $E_i^{p,0} \to E_\infty^{p,0}$.

Since $F_p^n = 0$ for $p > n$ we have $E_\infty^{p,0} \simeq F_p^p/F_{p+1}^p \simeq F_p^p$ so that one has an injective morphism $E_\infty^{p,0} \to H^p(K)$. Composing we have a morphism $E_i^{p,0} \to H^p(K)$. $\qquad\square$

The morphisms $H^q(K) \to E_i^{0,q}$ and $E_i^{p,0} \to H^p(K)$ are called *edge morphisms*.

The five term sequence relates some lower terms in a spectral sequence, and is often useful in practical calculations. We give here a standard proof of it; for more details and some other related sequences the reader may refer to [20, 64].

Proposition 5.9 (The five-term sequence). *Assume that $K_p^n = 0$ if $p > n$. There is an exact sequence*

$$0 \to E_2^{1,0} \to H^1(K) \to E_2^{0,1} \xrightarrow{d_2} E_2^{2,0} \to H^2(K). \tag{5.9}$$

Proof. For simplicity we only consider the case when the grading is over $\mathbb{N}$. The morphisms involved in the sequence in addition to d_2 have been defined in the previous two lemmas.

The injectivity of the morphism $E_2^{1,0} \to H^1(K)$ follows from the construction of Lemma 5.8, as the differential landing in $E_2^{1,0}$ starts from $E_2^{-1,1} = 0$.

As a second step we show that the kernel of d_2 coincides with the image of $H^1(K)$. Since the differential landing in $E_2^{0,1}$ starts from $E_2^{-2,2} = 0$, we have

$$E_3^{0,1} = \ker d_2 \colon E_2^{0,1} \to E_2^{2,0}. \tag{5.10}$$

We also note that at the third page of the spectral sequence the $(0,1)$ term has stabilized, since $d_i = 0$ on $E_i^{0,1}$ for $i \geq 3$. As a consequence, $E_3^{0,1} = \mathrm{Gr}(H^1(K))_0$, and there is an exact sequence

$$0 \to E_2^{1,0} \to H^1(K) \to E_3^{0,1} \to 0.$$

Splicing this sequence with (5.10) we obtain the exactness of (5.9) up to the fourth term.

To show the exactness at $E_2^{2,0}$ we note that $d_2 = 0$ on $E_2^{2,0}$, so that

$$E_3^{2,0} = \mathrm{coker}\, d_2 \colon E_2^{0,1} \to E_2^{2,0}.$$

Now, both the differentials landing in and starting from $E_i^{2,0}$ vanish for $i \geq 3$ so that the spectral sequence stabilizes at $E_3^{2,0}$, and

$$E_3^{2,0} = \mathrm{Gr}(H^2(K))_2 \subset H^2(K).$$

The kernel of the composition of the morphisms $E_2^{2,0} \to E_3^{2,0} \to H^2(K)$ coincides with the image of the differential landing in $E_2^{2,0}$, giving the desired exactness. $\qquad\square$

Exercise 5.10. Prove that the five-term sequence can be extended to an exact sequence with seven terms

$$0 \to E_2^{1,0} \to H^1(K) \to E_2^{0,1} \to E_2^{2,0} \to \ker e_2 \to E_2^{1,1} \xrightarrow{d_2} E_2^{3,0}, \qquad (5.11)$$

where e_2 is the edge morphism $H^2(K) \to E_2^{0,2}$.

Let us sketch how the morphism $\ker e_2 \to E_2^{1,1}$ is defined. Note that $\ker e_2 \simeq F_1^2$. Moreover, $E_2^{1,1}$ has no boundaries so that $E_3^{1,1} \subset E_2^{1,1}$. Finally, $E_\infty^{1,1} \simeq E_3^{1,1}$ as both differentials landing into $E_3^{1,1}$ and departing from it are zero. Then, the required morphism is defined by the composition

$$\ker e_2 = F_1^2 \to F_1^2/F_2^2 = E_\infty^{1,1} \simeq E_3^{1,1} \hookrightarrow E_2^{1,1}.$$

5.3. The Spectral Sequences Associated with a Double Complex

In this section, we consider a double complex as we have defined in Example 5.1.[c] As it was shown there, the total complex of a double complex has two natural filtrations, by columns and by rows, respectively. As a result, one has two spectral sequences, converging to the same group (however, the corresponding filtrations are in general different). Moreover, due to the presence of the bidegree, the result in Theorem 5.6 may be somehow refined.

We shall use the same notation as in Example 5.1; at first we consider the filtration by columns $T_\bullet$ of the total complex. The module

$$G = \bigoplus_{p \in \mathbb{Z}} T_p = \bigoplus_{p \in \mathbb{Z}} \bigoplus_{n \geq p, \; q \in \mathbb{N}} K^{n,q}$$

has a natural grading $G = \oplus_{n \in \mathbb{Z}} G^n$ given by

$$G^n = \bigoplus_{p \in \mathbb{Z}} T_p^n \simeq \bigoplus_{p \in \mathbb{Z}} \bigoplus_{j=0}^{n-p} K^{n-j,j}, \qquad (5.12)$$

---

[c] See, however, Remark 5.13.

but it also bigraded, with bidegree

$$G^{p,q} = T_q^{p+q}.$$

Note that if we form the total complex $\bigoplus_{p+q=n} G^{p,q}$ we obtain the complex (5.12) back:

$$\bigoplus_{p+q=n} G^{p,q} \simeq \bigoplus_{p+q=n} \bigoplus_{j=0}^{q} K^{p+q-j,j} = \bigoplus_{j=0}^{n-p} K^{n-j,j} = G^n.$$

The operators δ_1, δ_2 and $d = \delta_1 + \delta_2$ act on G:

$$\delta_1 \colon G^{p,q} \to G^{p+1,q}, \quad \delta_2 \colon G^{p,q} \to G^{p,q+1}, \quad d \colon G^n \to G^{n+1}.$$

We analyze the spectral sequence associated with these data. The first three terms are easily described. One has

$$E_0^{p,q} \simeq T_p^{p+q}/T_{p+1}^{p+q} \simeq K^{p,q}$$

while the differential $d_0 \colon E_0^{p,q} \to E_0^{p,q+1}$ coincides with $\delta_2 \colon K^{p,q} \to K^{p,q+1}$, and one has

$$E_1^{p,q} \simeq H^q(K^{p,\bullet}, \delta_2).$$

At next step we have $d_1 \colon E_1^{p,q} \to E_1^{p+1,q}$ with $E_1^{p,q} \simeq H^{p+q}(T_p/T_{p+1})$ and $T_p/T_{p+1} \simeq \bigoplus_{q \in \mathbb{Z}} K^{p,q}$. The differential

$$d_1 \colon H^{p+q}\left(\bigoplus_{n \in \mathbb{Z}} K^{p,n}\right) \to H^{p+q+1}\left(\bigoplus_{n \in \mathbb{Z}} K^{p+1,n}\right)$$

is identified with δ_1, and

$$E_2^{p,q} \simeq H^p(E_1^{\bullet,q}, \delta_1).$$

So far we have used the filtration by columns; if we instead use the filtration by rows, we obtain another spectral sequence, which we denote by $\{'E_i, 'd_i\}$. Both sequences converge to the same graded module, i.e., the cohomology of the total complex, but the corresponding filtrations are in general different, and this often provides interesting information. The second spectral sequence is formally obtained by replacing the complex $K^{p,q}$ with the complex $'K^{p,q} = K^{q,p}$. We get

$$'E_1^{p,q} \simeq H^q(K^{\bullet,p}, \delta_1),$$

$$'E_2^{p,q} \simeq H^p('E_1^{\bullet,q}, \delta_2).$$

Example 5.11. A simple application of the two spectral sequences associated with a double complex provides another proof of the de Rham Theorem

(Theorem 4.41), in the form of an isomorphism $\check{H}^\bullet(X, \mathbb{R}) \simeq H^\bullet_{dR}(X)$, where X is a differentiable manifold. Let $\mathfrak{U} = \{U_i\}$ be a good cover of X (see Section 4.4), and define the double complex

$$K^{p,q} = C^p(\mathfrak{U}, \Omega^q),$$

i.e., $K^{\bullet,q}$ is the complex of Čech cochains of $\mathfrak{U}$ with coefficients in the sheaf of differential q-forms. The first differential δ_1 is the Čech differential δ, while $\delta_2 = (-1)^p d$, where d is the exterior differential.[d] Note that with this choice of sign, the differentials δ_1 and δ_2 anticommute (this of course leaves the spaces of boundaries and cycles unchanged). We start analyzing the spectral sequences from the page E_1. For the first spectral sequence, we have

$$E_1^{p,q} \simeq H^q(K^{p,\bullet}, d) \simeq \prod_{i_0 < \cdots < i_p} H^q_{dR}(U_{i_0 \ldots i_p}).$$

Since all $U_{i_0 \ldots i_p}$ are contractible, we have

$$E_1^{p,0} \simeq C^p(\mathfrak{U}, \mathbb{R})$$

$$E_1^{p,q} = 0 \quad \text{for } q > 0.$$

As a consequence we have $E_2^{p,q} = 0$ for $q > 0$, while

$$E_2^{p,0} \simeq H^p(C^\bullet(\mathfrak{U}, \mathbb{R}), \delta) = H^p(\mathfrak{U}, \mathbb{R}).$$

This implies that $d_2 = 0$, so that the spectral sequence degenerates at the second page, and $E_\infty^{p,q} = 0$ for $q > 0$ and $E_\infty^{p,0} \simeq H^p(\mathfrak{U}, \mathbb{R})$. The resulting filtration of $H^p(T, D)$ has only one nonzero quotient, so that $H^p(T, D) \simeq H^p(\mathfrak{U}, \mathbb{R})$.

Let us now consider the second spectral sequence. We have

$$'E_1^{p,q} \simeq H^q(K^{\bullet,p}, \delta) = H^q(C^\bullet(\mathfrak{U}, \Omega^p), \delta) = H^q(\mathfrak{U}, \Omega^p).$$

Since the sheaves Ω^p are acyclic (see Proposition 4.38), we have

$$'E_1^{p,0} \simeq H^0(\mathfrak{U}, \Omega^p) \simeq \Omega^p(X)$$

$$'E_1^{p,q} = 0 \quad \text{for } q > 0.$$

At next step we have therefore $'E_2^{p,q} = 0$ for $q > 0$, and

$$'E_2^{p,0} = H^p('E_1^{\bullet,0}, d) \simeq H^p(\Omega^\bullet(X), d) = H^p_{dR}(X).$$

Again the spectral sequence degenerates at the second page, and we have $H^p(T, D) \simeq H^p_{dR}(X)$. Comparing with what we got from the first sequence, we obtain $H^p_{dR}(X) \simeq H^p(\mathfrak{U}, \mathbb{R})$. Taking a direct limit on good covers, we have $\check{H}^p(X, \mathbb{R}) \simeq H^p_{dR}(X)$.

---

[d]Here a notational conflict arises, so that we shall denote by D the differential of the total complex T.

Remark 5.12. As Example 5.11 shows, if at step r, with $r \geq 2$, one has $E_r^{p,q} = 0$ for $q > 0$ (or for $p > 0$ and $r \geq 1$), then the sequence degenerates at step r, and $E_r^{p,0} \simeq H^p(T^\bullet, d)$ (or $E_r^{0,q} \simeq H^q(T^\bullet, d)$). More generally, the same is true when E_r only has one row or one column; in this case $H^n(T^\bullet, d)$ is isomorphic to the only term $E_r^{p,q}$ with $p + q = n$.

Remark 5.13. Actually, according to equation (5.7), every spectral sequence can be regarded as a spectral sequence associated with a double complex.

Computing the d_2 differential. Computing the successive differentials of a spectral sequence is increasingly difficult. We have already understood the differentials d_0 and d_1; here we show how to compute d_2. We shall denote by $[\ ]_i$ the cohomology class in E_i of a cocycle in E_{i-1}.

Step 1. Let $\xi \in E_2^{p,q}$ and represent it as $[x]_2$, where $x \in E_1^{p,q}$ and $d_1 x = 0$. In turn, let $x = [b]_1$, where $b \in K^{p,q}$ and $\delta_2 b = 0$. Now, $\delta_1 b$ is a coboundary for δ_2, i.e., $\delta_1 b = -\delta_2 c$ for some $c \in K^{p+1,q-1}$, as

$$\delta_2 \delta_1 b = -\delta_1 \delta_2 b = 0,$$

and $[\delta_1 b]_1 = d_1 x = 0$. Summing up, ξ is represented by a class $b \in K^{p,q}$ such that

$$\delta_2 b = 0, \qquad \delta_1 b = -\delta_2 c.$$

Step 2. Note that

$$\delta_2 \delta_1 c = -\delta_1 \delta_2 c = -\delta_1 \delta_1 b = 0$$

so that we have a class $[\delta_1 c]_1 \in E_1^{p+2,q-1}$. Now we have

$$d_1 [\delta_1 c]_1 = [\delta_1^2 c]_1 = 0;$$

hence we may set $d_2 \xi = [\delta_1 c]_2 \in E_2^{p+2,q-1}$.

This "tic-tac-toe" construction is depicted in Figure 5.2.

5.4. Hypercohomology

Hypercohomology is a generalization of sheaf cohomology to the category of complexes of sheaves of $\mathcal{O}_X$-modules, i.e., it associates cohomology groups to a complex of $\mathcal{O}_X$-modules; more generally, to any left exact functor one can associate *hyperfunctors*, which act on complexes.

Definition 5.14. Given an abelian category $\mathfrak{A}$, a complex $A^\bullet$ of objects in $\mathfrak{A}$ is *bounded below* if $A^n = 0$ for all $n < n_0$ for some $n_0 \in \mathbb{Z}$.

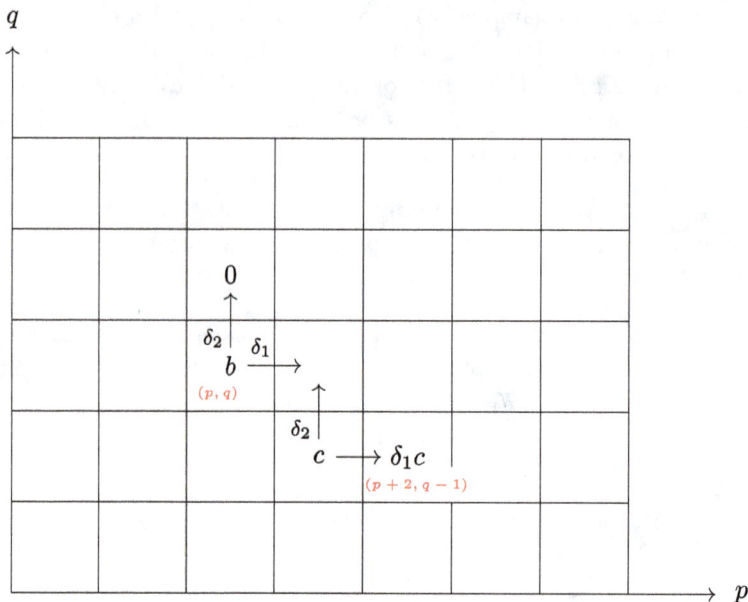

Figure 5.2.   Definition of the differential d_2.

We shall usually assume $n_0 = 0$.

Definition 5.15. A morphism of complexes $\phi \colon A^\bullet \to B^\bullet$ in an abelian category $\mathfrak{A}$ is a *quasi-isomorphism* if the morphisms induced in cohomology are isomorphisms.

One usually says that $B^\bullet$ is quasi-isomorphic to $A^\bullet$.

Lemma 5.16. *If $A^\bullet$ is a bounded below complex in an abelian category $\mathfrak{A}$ with enough injectives, there exists a quasi-isomorphism $f \colon A^\bullet \to I^\bullet$, where $I^\bullet$ is a complex of injective objects.*

Proof. We construct an injective resolution of $A^\bullet$, and then take its total complex. Embed A^0 into an injective object $I^{0,0}$, and form the diagram

$$
\begin{array}{ccccccccc}
& & & & & & 0 & & \\
& & & & & & \downarrow & & \\
0 & \longrightarrow & A^0 & \xrightarrow{\ \eta_0\ } & I^{0,0} \oplus A^1 & \longrightarrow & \operatorname{coker} \eta_0 & \longrightarrow & 0 \\
& & & & & & \downarrow & & \\
& & & & & & I^{1,0} & &
\end{array}
$$

where $I^{1,0}$ is injective; this produces a diagram

$$
\begin{array}{ccc}
0 & & 0 \\
\downarrow & & \downarrow \\
A^0 & \longrightarrow A^1 & \longrightarrow A^2 \\
\downarrow & & \downarrow \\
I^{0,0} & \xrightarrow{d_{0,0}} & I^{1,0}
\end{array}
$$

(note that $A^1 \to I^{1,0}$ is injective as A^1 injects into coker η_0). This procedure may be iterated: indeed we have a diagram

$$
\begin{array}{c}
0 \\
\downarrow \\
0 \longrightarrow A^1 \xrightarrow{\eta_1} I^{1,0} \oplus A^2 \longrightarrow \text{coker } \eta_1 \longrightarrow 0 \\
\downarrow \\
I^{2,0}
\end{array}
$$

and since $I^{2,0}$ is injective, there is a morphism $d_{1,0} \colon I^{1,0} \to I^{2,0}$, so that we get

$$
\begin{array}{ccc}
0 & 0 & 0 \\
\downarrow & \downarrow & \downarrow \\
A^0 \longrightarrow & A^1 \longrightarrow & A^2 \\
\downarrow & \downarrow & \downarrow \\
I^{0,0} \longrightarrow & I^{1,0} \longrightarrow & I^{2,0}
\end{array}
$$

and by iteration we obtain an injective morphism of complexes

$$
0 \to A^\bullet \to I^{\bullet,0}.
$$

Now, continuing similarly to Proposition 2.10, we obtain a double complex of injective objects $I^{\bullet,\bullet}$ which fits into a commutative diagram

$$
\begin{array}{ccccccc}
0 & & 0 & & 0 & & \\
\downarrow & & \downarrow & & \downarrow & & \\
A^0 & \longrightarrow & A^1 & \longrightarrow & A^2 & \longrightarrow & \cdots \\
\downarrow & & \downarrow & & \downarrow & & \\
I^{0,0} & \longrightarrow & I^{1,0} & \longrightarrow & I^{2,0} & \longrightarrow & \cdots \\
\downarrow & & \downarrow & & \downarrow & & \\
I^{0,1} & \longrightarrow & I^{1,1} & \longrightarrow & I^{2,1} & \longrightarrow & \cdots \\
\downarrow & & \downarrow & & \downarrow & & \\
I^{0,2} & \longrightarrow & I^{1,2} & \longrightarrow & I^{2,2} & \longrightarrow & \cdots \\
\downarrow & & \downarrow & & \downarrow & & \\
\vdots & & \vdots & & \vdots & &
\end{array}
$$

i.e., $I^{\bullet,\bullet}$ is an injective resolution of $A^\bullet$ (see Definition 5.23).

The reader can check that, after defining the total complex

$$I^i = \bigoplus_{p+q=i} I^{p,q},$$

there is a morphism of complexes $A^\bullet \to I^\bullet$ which is a quasi-isomorphism. $\qquad\square$

By Lemma 5.16, the following definition is well posed. Let $K_+(\mathfrak{A})$ be the category of bounded below complexes of objects in an abelian category $\mathfrak{A}$.

Definition 5.17. Consider a left exact functor of abelian categories $F\colon \mathfrak{A} \to \mathfrak{B}$, where $\mathfrak{A}$ has enough injectives. The right hyperderived functors $\mathbb{R}^n F\colon K_+(\mathfrak{A}) \to \mathfrak{B}$ are defined as

$$\mathbb{R}^n F(A^\bullet) = H^n(F(I^\bullet)),$$

where $I^\bullet$ is a complex of injective objects quasi-isomorphic to $A^\bullet$.

As a consequence of the following lemma, these hyperderived functors are independent, up to isomorphism, of the choice of the complex of injective objects.

Lemma 5.18. *If $\phi\colon A^\bullet \to B^\bullet$ is a morphism of complexes in an abelian category $\mathfrak{A}$, and $i\colon A^\bullet \to I^\bullet$, $j\colon B^\bullet \to J^\bullet$ are quasi-isomorphisms, with $I^\bullet$, $J^\bullet$ complexes of injectives, then there is a morphism $\tilde\phi\colon I^\bullet \to J^\bullet$, which lifts ϕ and is unique up to homotopy. As a consequence, for every left exact functor $F\colon \mathfrak{A} \to \mathfrak{B}$ there exist natural morphisms*

$$H^n(F(I^\bullet)) \to H^n(F(J^\bullet)).$$

Proof. A straightforward extension of the lifting property (Proposition 2.11) provides a morphism $\tilde\phi\colon I^\bullet \to J^\bullet$ making the diagram

$$
\begin{array}{ccc}
A^\bullet & \xrightarrow{\ i\ } & I^\bullet \\
\phi \downarrow & & \downarrow \tilde\phi \\
B^\bullet & \xrightarrow{\ j\ } & J^\bullet
\end{array}
$$

commutative. Two such lifts are always homotopic. Applying the functor F to $\tilde\phi$ and taking cohomology one obtains the required morphism. $\qquad\square$

Using this lemma for the morphism id$\colon A^\bullet \to A^\bullet$ and two complexes of injective objects quasi-isomorphic to $A^\bullet$, we obtain that the hyperderived functors do not depend, up to canonical isomorphism, on the choice of the quasi-isomorphic complex of injective objects.

As in the case of derived functors, the hyperderived functors can be computed via acyclic objects.

Definition 5.19. Given a left exact functor $F\colon \mathfrak{A} \to \mathfrak{B}$, a complex $A^\bullet$ in $\mathfrak{A}$ is F-acyclic if $R^n F(A^i) = 0$ for $n > 0$ and for every i.

Proposition 5.20 (Abstract de Rham Theorem for complexes).
Let $C^\bullet$ be an F-acyclic complex quasi-isomorphic to $A^\bullet$. There is a natural isomorphism

$$\mathbb{R}^n F(A^\bullet) \simeq H^n(F(C^\bullet)).$$

In other words, F-acyclic complexes compute the right hyperderived functors $\mathbb{R}F$.

Proof. The proof is analogous to that of Theorem 2.23. $\qquad\square$

Remark 5.21. If $A^\bullet$ is a complex concentrated in degree 0, i.e., $A^0 = A$ and $A^k = 0$ for all $k \neq 0$, then a complex of injective objects quasi-isomorphic to $A^\bullet$ is an injective resolution of A. Thus,

$$\mathbb{R}^n F(A^\bullet) \simeq R^n F(A).$$

In this sense, hyperderived functors generalize derived functors.

Example 5.22. Let $(X, \mathcal{O}_X)$ be a ringed space, and $\Gamma \colon \mathcal{O}_X\text{-}\mathbf{mod} \to \mathfrak{Ab}$ the global section functor. The right hyperderived functors $\mathbb{R}^n\Gamma$ of Γ are denoted by $\mathbb{H}^n(X, -)$ and are called the *hypercohomology* of $(X, \mathcal{O}_X)$ with coefficients in a complex of $\mathcal{O}_X$-modules.

5.5. The Spectral Sequence of Hyperderived Functors

A key feature of right hyperderived functors is that there exists a spectral sequence associated to them. The study of this spectral sequence is facilitated by the use of a special class of injective resolutions.

5.5.1. Cartan–Eilenberg resolutions

The Horseshoe Lemma 2.18 can be used to construct injective resolutions that are well suited to deal with derived functors (in particular we shall use them in Section 5.5.2 to study hyperderived functors) [64]. Let $\mathfrak{A}$ be an abelian category with enough injectives, and let $A^\bullet$ be a bounded below complex of objects in $\mathfrak{A}$.

Definition 5.23. A resolution of $A^\bullet$ is a double complex $L^{\bullet,\bullet}$ such that

- for every p the complex $L^{p,\bullet}$ is a resolution of A^p;
- all diagrams

$$
\begin{array}{ccccc}
0 & \longrightarrow & A^p & \longrightarrow & I^{p,\bullet} \\
 & & \downarrow & & \downarrow \\
0 & \longrightarrow & A^{p+1} & \longrightarrow & I^{p+1,\bullet}
\end{array}
$$

commute.

Proposition 5.24. *There exist injective resolutions $I^{\bullet,\bullet}$ of $A^\bullet$ such that for every p the complex $H^p(I^{\bullet,\bullet})$ (cohomology taken with respect to the first index) is a resolution of the cohomology object $H^p(A^\bullet)$.*

Proof. Denoting by $B^p(A)$ and $Z^p(A)$ the coboundary and cocycle subobjects of A^p, for every p there is an exact sequence

$$0 \to B^p(A) \to Z^p(A) \to H^p(A^\bullet) \to 0.$$

Applying the Horseshoe Lemma we obtain injective resolutions $I^\bullet_{B,p}$, $I^\bullet_{Z,p}$, $I^\bullet_{H,p}$ of $B^p(A)$, $Z^p(A)$, $H^p(A^\bullet)$, respectively, fitting into a diagram

$$
\begin{array}{ccccccccc}
0 & \longrightarrow & B^p(A) & \longrightarrow & Z^p(A) & \longrightarrow & H^p(A^\bullet) & \longrightarrow & 0 \\
& & \downarrow & & \downarrow & & \downarrow & & \\
0 & \longrightarrow & I^\bullet_{B,p} & \longrightarrow & I^\bullet_{Z,p} & \longrightarrow & I^\bullet_{H,p} & \longrightarrow & 0
\end{array}
\tag{5.13}
$$

On the other hand, one has exact sequences

$$0 \to Z^p(A) \to A^p \to B^{p+1}(A) \to 0$$

and again using the Horseshoe Lemma one can find injective resolutions $I^\bullet_{A,p}$ of the objects A^p fitting into a diagram

$$
\begin{array}{ccccccccc}
0 & \longrightarrow & Z^p(A) & \longrightarrow & A^p & \longrightarrow & B^{p+1}(A) & \longrightarrow & 0 \\
& & \downarrow & & \downarrow & & \downarrow & & \\
0 & \longrightarrow & I^\bullet_{Z,p} & \longrightarrow & I^\bullet_{A,p} & \longrightarrow & I^\bullet_{B,p+1} & \longrightarrow & 0
\end{array}
$$

where the resolutions $I^\bullet_{Z,p}$ and $I^\bullet_{B,p}$ are the same as in equation (5.13). Now one defines a double complex

$$I^{p,q} = I^q_{A,p}$$

whose vertical arrows are the differentials of $I^\bullet_{A,p}$ multiplied by $(-1)^p$, while the horizontal arrows are given by the compositions

$$I^{p,q} = I^q_{A,p} \to I^q_{B,p+1} \to I^q_{Z,p+1} \to I^q_{A,p+1} = I^{p+1,q}.$$

One can check that:

(1) $I^{\bullet,\bullet}$ is a resolution of $A^\bullet$;
(2) the cohomology $H^p(I^{\bullet,\bullet})$ of $I^{\bullet,\bullet}$ with respect to the first index (i.e., the cohomology of the rows of the double complex), regarded as a complex in the second index, is isomorphic to $I^\bullet_{H,p}$ (which is an injective resolution of $H^p(A^\bullet)$). $\quad\square$

$I^{\bullet,\bullet}$ is a *Cartan–Eilenberg resolution* of $A^\bullet$. One should note that these resolutions also have the following evident properties:

(1) if $A^p = 0$, then $I^{p,\bullet} = 0$;
(2) the injective resolutions $I^\bullet_{B,p}$ of $B^p(A)$ are the images of the morphisms $I^{p,\bullet} \to I^{p+1,\bullet}$.

For more details the reader may refer to [64].

5.5.2. The hyperderived functors spectral sequence

We construct now the spectral sequence attached to a hyperderived functor. This will have as a special case the Grothendieck spectral sequence associated with a composition of functors.

Let $(A^\bullet, f)$ be a bounded below complex in an abelian category $\mathfrak{A}$ with enough injectives, and take an injective resolution $0 \to A^\bullet \to I^{\bullet,\bullet}$, as in Lemma 5.16. Given a left exact functor $F \colon \mathfrak{A} \to \mathfrak{B}$, we consider the double complex

$$K^{p,q} = F(I^{p,q})$$

$$K^{p+1,q} \overset{f}{\swarrow} \qquad \overset{d}{\searrow} K^{p,q+1}$$

where f is induced by $f \colon A^\bullet \to A^{\bullet+1}$ and d is induced by the differential of the injective resolution (that is, we are denoting $F(f)$ by f, and $F(d)$ by d).

The first pages of the first spectral sequence, i.e., the one induced by the filtration by columns, are given by

$$E_1^{p,q} = H^q(F(I^{p,\bullet}), d) = R^q F(A^p), \tag{5.14}$$

$$E_2^{p,q} = H^p(E_1^{\bullet,q}, f) = H^p(R^q F(A^\bullet), f), \tag{5.15}$$

while the first pages of the second spectral sequence are

$$'E_1^{p,q} = H^q(F(I^{\bullet,p}), f), \tag{5.16}$$

$$'E_2^{p,q} = H^p(H^q(F(I^{\bullet,\bullet})), d) \simeq R^p F(H^q(A^\bullet)). \tag{5.17}$$

Here $H^q(F(I^{\bullet,\bullet}))$ is the cohomology with respect to the first index, and H^p with respect to the second. The last isomorphism is proved by choosing for $I^{\bullet,\bullet}$ a Cartan–Eilenberg resolution (see Proposition 5.24), so that, for each q, the objects $H^q(I^{\bullet,p})$ make up an injective resolution of $H^q(A^\bullet)$. As a consequence, the cohomology of the complex $F(H^q(I^{\bullet,p}))$ is $R^p F(H^q(A^\bullet))$.

On the other hand, since the total complex of $I^{\bullet,\bullet}$ is quasi-isomorphic to $A^\bullet$ (see proof of Lemma 5.16), the cohomology of the total complex of $K^{\bullet,\bullet}$ is given by the hyperderived functors $\mathbb{R}^p F(A^\bullet)$.

The next theorem resumes what we have so far found about the spectral sequences associated to hyperderived functors.

Theorem 5.25. *Let $F \colon \mathfrak{A} \to \mathfrak{B}$ be a left exact functor between abelian categories, $\mathfrak{A}$ with enough injectives. Let $A^\bullet$ be a bounded below complex* *in $\mathfrak{A}$. There are two spectral sequences (E_i, d_i) and $('E_i, 'd_i)$, both converging to $\mathbb{R}F(A^\bullet)$, whose first pages are given by equations (5.14), (5.15), (5.16) and (5.17).*

We give now two important applications of theses spectral sequences. In the first we consider a complex $L^\bullet$ in $\mathfrak{A}$ which is a resolution of an object A. In this case for the first spectral sequence we get

$$E_1^{p,q} = R^q F(L^p), \tag{5.18}$$

$$E_2^{p,q} = H^p(R^q F(L^\bullet)), \tag{5.19}$$

while the only nonzero terms of the second page of the second spectral sequence are

$$'E_2^{p,0} = R^p F(A),$$

so that this spectral sequence degenerates at the second page and converges to $RF(A)$. Therefore we have proved the following theorem.

Theorem 5.26. *Let $F \colon \mathfrak{A} \to \mathfrak{B}$ be a left exact functor between abelian categories, and assume that $\mathfrak{A}$ has enough injectives. If $L^\bullet$ is a resolution of an object A in $\mathfrak{A}$ there is a spectral sequence, whose first two pages are given in equations (5.18) and (5.19), which converges to $RF(A)$.*

Example 5.27. Let $\mathfrak{U}$ be an open cover of a topological space X, and $\mathcal{F}$ a sheaf of abelian groups on X. If we apply Theorem 5.26 to the Čech resolution $\check{\mathcal{C}}^\bullet(\mathfrak{U}, \mathcal{F})$ of $\mathcal{F}$ (see Section 4.3), taking for F the global section functor, we obtain a spectral sequence converging to the sheaf cohomology $H^\bullet(X, \mathcal{F})$ of $\mathcal{F}$ whose first page is

$$E_1^{p,q} = H^q(X, \check{\mathcal{C}}^p(\mathfrak{U}, \mathcal{F})).$$

A corollary to Theorem 5.25 is an important result due to Grothendieck [22, Theorem 2.4.1]. Many facts in homological algebra can be proved in a unified way using this result.

Theorem 5.28 (Grothendieck's spectral sequence). *Consider a composition of functors*

$$\mathfrak{A} \xrightarrow{F} \mathfrak{B} \xrightarrow{G} \mathfrak{C},$$

where $\mathfrak{A}$, $\mathfrak{B}$ and $\mathfrak{C}$ are abelian categories, $\mathfrak{A}$ and $\mathfrak{B}$ have enough injectives, F and G are left exact functors. Suppose that F maps injective objects of $\mathfrak{A}$ to G-acyclic objects of $\mathfrak{B}$. Then, for every object A in $\mathfrak{A}$, there exists a spectral sequence $('E_i, d_i)$ converging to $R(G \circ F)(A)$, whose second page is

$$'E_2^{p,q} = R^p G(R^q F(A)). \tag{5.20}$$

Proof. Consider an injective resolution of A, say $0 \to A \to I^\bullet$. We apply Theorem 5.25 to the functor G and the complex $F(I^\bullet)$; the second page of the first spectral sequence is

$$E_2^{p,q} = H^p(R^q G(F(I^\bullet))).$$

Since the terms of the complex $F(I^\bullet)$ are G-acyclic, the only nonzero terms of E_2 are

$$E_2^{p,0} = H^p(G(F(I^\bullet))) = R^p(G \circ F)(A),$$

so that this spectral sequence degenerates and converges to $R(G \circ F)(A)$.

On the other hand, the second page of the second spectral sequence is that in equation (5.20), see equation (5.17): indeed, $H^q(F(I^\bullet)) = R^q F(A)$.   □

Analogous spectral sequences can be associated with compositions of functors of different nature. For instance, let $\mathfrak{A}$, $\mathfrak{B}$, $\mathfrak{C}$ be abelian categories, where $\mathfrak{A}$ has enough projectives and $\mathfrak{B}$ has enough injectives.

Theorem 5.29. *Let $F \colon \mathfrak{A} \to \mathfrak{B}$ be a right exact functor which maps projective objects of $\mathfrak{A}$ to G-acyclic objects of $\mathfrak{B}$, where $G \colon \mathfrak{B} \to \mathfrak{C}$ is a contravariant left exact functor. For each object $A \in \mathfrak{A}$, there is a spectral sequence converging to $R(G \circ F)(A)$, whose second page is*

$$E_2^{p,q} = R^p G(L_q F(A)).$$

5.6. Some Applications

We are now going to describe some concrete spectral sequences. In particular, we shall analyze the local-to-global sequence (which relates the local and global Ext functors), the Leray sequence (which relates the cohomology of a sheaf on the total space of a fibration with its higher direct images), and the Hochschild–Serre sequence associated with an extension of Lie algebras. Moreover, we shall discuss the spectral sequence associated with an open cover (Čech spectral sequence), and some spectral sequences of the Künneth type.

5.6.1. Deriving Hom in the first variable

As we anticipated in Section 2.3, one can realize the Ext functors in an abelian category $\mathfrak{A}$ by deriving the Hom functor with respect to its first argument (while they were originally defined by deriving with respect to the second argument). This requires that the category $\mathfrak{A}$ has both enough injectives and projectives, as it happens, for instance, for the category R-**mod**, with R a commutative ring with unity. In this section, we offer a proof of this fact based on the spectral sequences associated with a suitable double complex.

Proposition 5.30. *Let $\mathfrak{A}$ be an abelian category with both enough injectives and projectives. For every object b in $\mathfrak{A}$, the right derived functors of the left exact functor*

$$\mathrm{Hom}_{\mathfrak{A}}(-, b) \colon \mathfrak{A} \to \mathfrak{Ab}$$

are isomorphic to the functors $\mathrm{Ext}^i_{\mathfrak{A}}(-, b)$.

We shall actually prove that, for any object $a \in \mathfrak{A}$, there is an isomorphism $\mathrm{Ext}^p_{\mathfrak{A}}(a, b) \simeq H^p(\mathrm{Hom}_{\mathfrak{A}}(P_\bullet, b))$, where $P_\bullet$ is a projective resolution of a. This means that we may regard the *contravariant* functor $\mathrm{Hom}_{\mathfrak{A}}(-, b) \colon \mathfrak{A} \to \mathfrak{Ab}$ as a covariant functor $\mathfrak{A}^{\mathrm{op}} \to \mathfrak{Ab}$, deriving it using injective resolutions in $\mathfrak{A}^{\mathrm{op}}$, which are projective resolutions in $\mathfrak{A}$.

Proof. Let $P_\bullet \to a \to 0$ be a projective resolution of an object $a \in \mathfrak{A}$, and let $0 \to b \to I^\bullet$ be an injective resolution of b. Form the double complex

$$K^{p,q} = \mathrm{Hom}_{\mathfrak{A}}(P_p, I^q).$$

We first analyze the second spectral sequence, whose first page is

$$'E_1^{p,q} = H^q(K^{\bullet, p}) = H^q(\mathrm{Hom}_{\mathfrak{A}}(P_\bullet, I^p)).$$

Since the objects P_i are projective, the functors $\mathrm{Hom}_{\mathfrak{A}}(P_i, -)$ are exact (see Section 2.7), so that they commute with taking cohomology, and we can write

$$'E_1^{p,q} \simeq \mathrm{Hom}_{\mathfrak{A}}(H_q(P_\bullet), I^p).$$

Since $P_\bullet$ is a resolution of a, we have

$$'E_1^{p,q} \simeq \begin{cases} \mathrm{Hom}_{\mathfrak{A}}(a, I^p) & \text{for } q = 0, \\ 0 & \text{for } q > 0, \end{cases}$$

so that

$$'E_2^{p,q} \simeq \begin{cases} H^p(\mathrm{Hom}_{\mathfrak{A}}(a, I^\bullet)) = \mathrm{Ext}^p_{\mathfrak{A}}(a, b) & \text{for } q = 0, \\ 0 & \text{for } q > 0. \end{cases}$$

Thus, this spectral sequence degenerates at the second page and converges to the groups $\mathrm{Ext}^p_{\mathfrak{A}}(a, b)$.

Now we examine the first spectral sequence; one has

$$E_1^{p,q} = H^q(K^{p,\bullet}) = H^q(\mathrm{Hom}_{\mathfrak{A}}(P_p, I^{\bullet})).$$

Again, since the objects P_i are projective, and $I^{\bullet}$ is a resolution of b, for every p the complex $\mathrm{Hom}_{\mathfrak{A}}(P_p, I^{\bullet})$ is exact in positive degree and $H^0(\mathrm{Hom}_{\mathfrak{A}}(P_p, I^{\bullet})) = \mathrm{Hom}_{\mathfrak{A}}(P_p, b)$, i.e.,

$$E_1^{p,q} = \begin{cases} \mathrm{Hom}_{\mathfrak{A}}(P_p, b) & \text{for } q = 0, \\ 0 & \text{for } q > 0, \end{cases}$$

and then

$$E_2^{p,q} = \begin{cases} H^p(\mathrm{Hom}_{\mathfrak{A}}(P_\bullet, b)) & \text{for } q = 0, \\ 0 & \text{for } q > 0. \end{cases}$$

So also this spectral sequence degenerates at the second page, and comparing the two spectral sequences we obtain

$$\mathrm{Ext}_{\mathfrak{A}}^p(a, b) \simeq H^p(\mathrm{Hom}_{\mathfrak{A}}(P_\bullet, b)). \qquad \square$$

5.6.2. The local-to-global spectral sequence

Let $(X, \mathcal{O}_X)$ be a ringed space. One can define a spectral sequence which relates the local and global Ext functors $\mathcal{E}xt_{\mathcal{O}_X}$ and $\mathrm{Ext}_{\mathcal{O}_X}$. We recall that these are the right derived functors of the local Hom functor

$$\mathcal{H}om_{\mathcal{O}_X}(\mathcal{E}, -) \colon \mathcal{O}_X\text{-mod} \to \mathcal{O}_X\text{-mod}$$

and of the global Hom functor

$$\mathrm{Hom}_{\mathcal{O}_X}(\mathcal{E}, -) \colon \mathcal{O}_X\text{-mod} \to \mathfrak{Ab},$$

respectively.

Theorem 5.31 (Local-to-global spectral sequence). *Let $(X, \mathcal{O}_X)$ be a ringed space and let $\mathcal{E}$ and $\mathcal{F}$ be $\mathcal{O}_X$-modules. There is a spectral sequence converging to $\mathrm{Ext}_{\mathcal{O}_X}^\bullet(\mathcal{E}, \mathcal{F})$ whose second page is*

$$E_2^{p,q} = H^p(X, \mathcal{E}xt_{\mathcal{O}_X}^q(\mathcal{E}, \mathcal{F})).$$

Proof. The local and global Ext functors are related by

$$\mathrm{Hom}_{\mathcal{O}_X}(\mathcal{E}, -) = \Gamma \circ \mathcal{H}om_{\mathcal{O}_X}(\mathcal{E}, -).$$

In view of Theorem 5.28, we only need to show that the functor $\mathcal{H}om_{\mathcal{O}_X}(\mathcal{E}, -)$ maps injective $\mathcal{O}_X$-modules to Γ-acyclic $\mathcal{O}_X$-modules. In particular, one shows that if $\mathcal{I}$ is an injective $\mathcal{O}_X$-module, then

$\mathcal{H}om_{\mathcal{O}_X}(\mathcal{E}, \mathcal{I})$ is flabby, so that it is Γ-acyclic by Theorem 4.22. Let U be an open subset in X, and $s \in \mathcal{H}om_{\mathcal{O}_X}(\mathcal{E}, \mathcal{I})(U)$; then s may be regarded as morphism $s \colon \mathcal{E}_U = j_!(\mathcal{E}_{|U}) \to \mathcal{I}$ (the functor $j_!$ was defined in equation (4.9)). Since $\mathcal{E}_U$ is an $\mathcal{O}_X$-submodule of $\mathcal{E}$, and $\mathcal{I}$ is injective, s extends to a morphism $\mathcal{E} \to \mathcal{I}$, which is the required global section of $\mathcal{H}om_{\mathcal{O}_X}(\mathcal{E}, \mathcal{I})$.

Note that the spectral sequence of Theorem 5.28 in this case converges to

$$R^n \mathrm{Hom}_{\mathcal{O}_X}(\mathcal{E}, \mathcal{F}) = \mathrm{Ext}^n_{\mathcal{O}_X}(\mathcal{E}, \mathcal{F}),$$

and the second page is

$$E_2^{p,q} = R^p \Gamma(X, R^q \mathcal{H}om_{\mathcal{O}_X}(\mathcal{E}, \mathcal{F})) = H^p(X, \mathcal{E}xt^q_{\mathcal{O}_X}(\mathcal{E}, \mathcal{F})). \qquad \square$$

5.6.3. The Čech spectral sequence

Let $\mathcal{F}$ be a sheaf of abelian groups on X and let $\mathfrak{U}$ be an open cover of X. One can construct a spectral sequence which relates the Čech cohomology of $\mathfrak{U}$ with coefficients in $\mathcal{F}$ with the sheaf cohomology of $\mathcal{F}$. Pick up an injective resolution $\mathcal{I}^\bullet$ of $\mathcal{F}$, and form the double complex

$$K^{p,q} = C^p(\mathfrak{U}, \mathcal{I}^q); \tag{5.21}$$

the two differentials are, up to signs, the Čech differential δ and the differential δ_2 of the complex $\mathcal{I}^\bullet$, which commute (see footnote e in Chapter 2).

First we study the spectral sequence associated with the filtration by rows. The first page is

$$'E_1^{p,q} = H^q(C^\bullet(\mathfrak{U}, \mathcal{I}^p)) = H^q(\mathfrak{U}, I^p).$$

Since the sheaves $\mathcal{I}^p$ are flabby, we have $'E_1^{p,q} = 0$ for $q > 0$ (cf. Lemma 4.29), and $'E_1^{p0} = \Gamma(X, \mathcal{I}^p)$, so that $'E_2^{p,q} = 0$ for $q > 0$, and

$$'E_2^{p0} = H^p(X, \mathcal{F}) \simeq \,'E_\infty^{p0}.$$

Therefore, the spectral sequence degenerates at the second page and converges to the sheaf cohomology of $\mathcal{F}$.

Then, we consider the spectral sequence induced by the filtration by columns. The first page is

$$E_1^{p,q} = H^q(C^p(\mathfrak{U}, \mathcal{I}^\bullet)) \simeq \prod_{i_0 < \cdots < i_p} H^q(U_{i_0,\dots,i_p}, \mathcal{F}). \tag{5.22}$$

We can use this spectral sequence to yield a slightly different proof of the Leray Theorem 4.33. Indeed, if the condition

$$H^q(U_{i_0\ldots i_p}, \mathcal{F}) = 0 \qquad (5.23)$$

holds for all $q > 0$ and all nonvoid intersections $U_{i_0\ldots i_p}$, the first page of the spectral sequence is (cf. Proposition 4.26 (1))

$$E_1^{p,q} = \begin{cases} \displaystyle\prod_{i_0 < \cdots < i_p} \mathcal{F}(U_{i_0\cdots i_p}) \simeq C^p(\mathfrak{U}, \mathcal{F}) & \text{for } q = 0, \\ 0 & \text{for } q > 0. \end{cases}$$

Since $d_1 \colon E_1^{p,q} \to E_1^{p+1,q}$, the differential is nonzero only when acting on $E_1^{p,0}$, so that (as then d_1 is the Čech differential)

$$E_2^{p,q} = \begin{cases} H^p(\mathfrak{U}, \mathcal{F}) & \text{for } q = 0, \\ 0 & \text{for } q > 0. \end{cases}$$

Now since $d_2 \colon E_2^{p,q} \to E_2^{p+2,q-1}$ we have $d_2 = 0$, so that the spectral sequence degenerates at E_2, and its limit is $E_2^{p,0}$. On the other hand, the limit is also $H^\bullet(X, \mathcal{F})$, so that

$$H^p(\mathfrak{U}, \mathcal{F}) \simeq H^p(X, \mathcal{F}) \quad \text{for } p \geq 0.$$

One can compute the second page E_2 even when the condition (5.23) does not hold. To this end, let us define the presheaf

$$\mathcal{H}^q(\mathcal{F})(U) = H^q(U, \mathcal{F})$$

for $q \geq 0$. Note that $\mathcal{H}^0(\mathcal{F}) \simeq \mathcal{F}$. The first page E_1 can be rewritten

$$E_1^{p,q} = C^p(\mathfrak{U}, \mathcal{H}^q(\mathcal{F}))). \qquad (5.24)$$

Since d_1 is the Čech differential, we have

$$E_2^{p,q} \simeq H^p(\mathfrak{U}, \mathcal{H}^q(\mathcal{F})). \qquad (5.25)$$

We have therefore the following fundamental result.

Theorem 5.32 (Čech spectral sequence). *Let $\mathfrak{U}$ be an open cover of a topological space X, and let $\mathcal{F}$ be a sheaf of abelian groups on X. There is a spectral sequence whose first and second pages are given in equations (5.22) (or (5.24)) and (5.25), respectively, and converges to the sheaf cohomology groups $H^\bullet(X, \mathcal{F})$.*

For future use, we can make the five-term sequence (5.9) of this spectral sequence explicit:

$$0 \to H^1(\mathfrak{U}, \mathcal{F}) \to H^1(X, \mathcal{F}) \to H^0(\mathfrak{U}, \mathcal{H}^1(\mathcal{F}))$$
$$\to H^2(\mathfrak{U}, \mathcal{F}) \to H^2(X, \mathcal{F}). \tag{5.26}$$

Now one has [22, Corollary 2, p. 176]:

Lemma 5.33. $\check{H}^0(X, \mathcal{H}^q(\mathcal{F})) = 0$ *for every* $q > 0$.

Proof. We follow the proof in [22]. The lemma is a consequence of the following fact: if U is an open neighbourhood of $x \in X$, and $\xi \in H^q(U, \mathcal{F})$, there exists an open subset $V \subset U$ containing x such that ξ restricts to zero on $H^q(V, \mathcal{F})$. We prove this statement. Let $\mathcal{I}^\bullet$ be an injective resolution of $\mathcal{F}|_U$, and represent ξ with a cocycle $z \in \mathcal{I}^q(U)$. Since the complex $\mathcal{I}^\bullet$ is exact on the stalks, there is an open neighbourhood V of X such that $z|_V$ is a coboundary. This proves the claim. □

If we take the direct limit of the exact sequence (5.26) over the open cover $\mathfrak{U}$, in view of Lemma 5.33 one obtains that the morphism $\check{H}^1(X, \mathcal{F}) \to H^1(X, \mathcal{F})$ is an isomorphism, while $\check{H}^2(X, \mathcal{F}) \to H^2(X, \mathcal{F})$ is injective, confirming what we already found in Remark 4.32.

These results can be used to complete the Example 4.28, i.e., an example of a sheaf $\mathcal{F}$ for which the morphism $\check{H}^2(X, \mathcal{F}) \to H^2(X, \mathcal{F})$ is not an isomorphism. To this end, we need a companion to Lemma 5.33.

Lemma 5.34. *If X is a topological space, and $\mathcal{F}$ is a sheaf on X, there is an exact sequence*

$$0 \to \check{H}^2(X, \mathcal{F}) \to H^2(X, \mathcal{F}) \to \check{H}^1(X, \mathcal{H}^1(\mathcal{F})) \to 0. \tag{5.27}$$

Proof. In view of Lemma 5.33, the seven-term sequence (5.11) yields a segment

$$0 \to H^2(\mathfrak{U}, \mathcal{F}) \to H^2(X, \mathcal{F}) \to H^1(\mathfrak{U}, \mathcal{H}^1(\mathcal{F})) \xrightarrow{d_2} H^3(\mathfrak{U}, \mathcal{F}).$$

To prove the lemma it will be sufficient to show that $d_2 = 0$ when the open cover $\mathfrak{U}$ is fine enough. So we compute d_2 (see Section 5.3, p. 139). If we represent $\xi \in E_2^{1,1}$ with an element

$$b \in K^{1,1} = \prod_{i<j} \mathcal{I}^1(U_{ij})$$

such that $\delta_2 b = 0$ then $d_2(\xi) = [\delta c]$, where

$$c \in K^{2,0} = \prod_{i<j<k} \mathcal{I}^0(U_{ijk})$$

is such that $\delta b = -\delta_2 c$. Now, if the open cover $\mathfrak{U}$ is fine enough, the sequences

$$\mathcal{I}^0(U_{ij}) \xrightarrow{\delta_2} \mathcal{I}^1(U_{ij}) \xrightarrow{\delta_2} \mathcal{I}^2(U_{ij})$$

are exact, so that $b = \delta_2 b'$ for $b' \in K^{1,0}$, and $\delta c = -\delta^2 b' = 0$. $\qquad\square$

Example 5.35. We are now in a position to complete Example 4.28. We shall again follow [22]. Let $\mathcal{F} = j_!(\Bbbk_{X-Y})$; we already proved that $H^2(X, \mathcal{F}) = \Bbbk$. We shall now prove that $\check{H}^1(X, \mathcal{H}^1(\mathcal{F})) = \Bbbk$; in view of the exact sequence (5.27), this will imply $\check{H}^2(X, \mathcal{F}) = 0$.

Step 1. We compute the sheaf $\mathcal{H}^1(\mathcal{F})$. Let $V \subset X$ be an open subset, and set $Y' = V \cap Y$. The analogue of the sequence (4.12) for the pair (V, Y') yields the exact sequence

$$H^0(V, \Bbbk_V) \to H^0(Y', \Bbbk_{Y'}) \to H^1(V, \mathcal{F}) \to 0.$$

The quotient $H^1(V, \mathcal{F})$ is nonzero only when Y' has two connected components, which happens when V meets both Y_1 and Y_2 but not their intersection. So $\mathcal{H}^1(\mathcal{F})(V) \neq 0$ only in this special case.

Step 2. We compute

$$\check{H}^1(X, \mathcal{H}^1(\mathcal{F})) = \varinjlim_{\mathfrak{U}} H^1(\mathfrak{U}, \mathcal{H}^1(\mathcal{F})).$$

We consider open covers $\mathfrak{U} = \{U_x\}_{x \in X}$, where each U_x only meets one of the closed subsets Y_1 and Y_2 unless x is one of the intersections point of $Y_1 \cap Y_2$, in which case we assume that U_x does not contain the other. It is easy to check that this implies $C^0(\mathfrak{U}, \mathcal{H}^1(\mathcal{F})) = 0$, i.e., no U_x is of the "special type" of the previous step. Thus, $H^1(\mathfrak{U}, \mathcal{H}^1(\mathcal{F})) = Z^1(\mathfrak{U}, \mathcal{H}^1(\mathcal{F}))$. On the other hand, if $x, y \in X$ then $\mathcal{H}^1(\mathcal{F})(U_x \cap U_y) = 0$, unless x, y are the intersection points of Y_1 and Y_2 in which case $U_x \cap U_y$ is of "special type". Therefore,

$$Z^1(\mathfrak{U}, \mathcal{H}^1(\mathcal{F})) = C^1(\mathfrak{U}, \mathcal{H}^1(\mathcal{F})) = \Bbbk.$$

So $H^1(\mathfrak{U}, \mathcal{H}^1(\mathcal{F})) = \Bbbk$, and taking a direct limit $\check{H}^1(X, \mathcal{H}^1(\mathcal{F})) = \Bbbk$.

5.6.4. The Čech spectral sequence for complexes

The Čech spectral sequence of Theorem 5.32 can be extended to complexes of sheaves. For any double complex $K^{\bullet,\bullet}$ denote by $\mathrm{Tot}^\bullet(K)$ the associate total complex, i.e.,

$$\mathrm{Tot}^n(K) = \bigoplus_{p+q=n} K^{p,q}$$

with the usual differential (see footnote e in Chapter 2).

Theorem 5.36. *Let $\mathfrak{U}$ be an open cover of a topological space X, and let $\mathcal{F}^\bullet$ be a bounded below complex of sheaves of abelian groups on X. There is a spectral sequence converging to the hypercohomology groups $\mathbb{H}^\bullet(X, \mathcal{F}^\bullet)$ whose first pages are*

$$E_1^{p,q} = \bigoplus_{n+m=p} C^n(\mathfrak{U}, \mathcal{H}^q(\mathcal{F}^m)), \quad E_2^{p,q} = H^p(\mathrm{Tot}^\bullet(C(\mathfrak{U}, \mathcal{H}^q(\mathcal{F})))).$$

This is the spectral sequence associated to the filtration by columns of the double complex (5.21),

$$K^{p,q} = C^p(\mathfrak{U}, \mathcal{I}^q)$$

where $\mathcal{I}^\bullet$ is a complex of injective sheaves quasi-isomorphic to $\mathcal{F}^\bullet$.

For a proof of Theorem 5.36 the reader may for instance refer to [49, Lemma 20.26.1]. One can easily check that for a concentrated complex $\mathcal{F}$, Theorem 5.36 reduces to Theorem 5.32, by applying it to an injective resolution of $\mathcal{F}$.

5.6.5. The Leray spectral sequence

As we saw in Section 3.4, given a continuous map of topological spaces $f: X \to Y$, the direct image functor $f_*: \mathfrak{Sh}_X \to \mathfrak{Sh}_Y$ is additive and left exact, so that one can consider its right derived functors $R^i f_*: \mathfrak{Sh}_X \to \mathfrak{Sh}_Y$, the higher direct images of f. This construction gives rise to a spectral sequence.

Theorem 5.37. *Let $f: X \to Y$ be a continuous map of topological spaces. Then, for every $\mathcal{F} \in \mathfrak{Sh}_X$, there exists a spectral sequence converging to $H^\bullet(X, \mathcal{F})$, whose second page is*

$$E_2^{p,q} = H^p(Y, R^q f_* \mathcal{F}).$$

Proof. Consider the composition of functors

$$\mathfrak{Sh}_X \xrightarrow{f_*} \mathfrak{Sh}_Y \xrightarrow{\Gamma_Y} \mathfrak{Ab}.$$

Now, f_* maps injective objects to Γ_Y-acyclic objects; indeed, injective sheaves of abelian groups are flabby (Remark 4.21), while by its very definition, f_* maps flabby sheaves to flabby sheaves, and these are Γ-acyclic (Theorem 4.22). Note that the composition is nothing but the global section functor of X, i.e., $(\Gamma_Y \circ f_*)(\mathcal{F}) = \Gamma_X(\mathcal{F})$. As a consequence, by Grothendieck's Theorem 5.28, there is a spectral sequence converging to $R^\bullet \Gamma_X(\mathcal{F}) = H^\bullet(X, \mathcal{F})$, whose second page is

$$E_2^{p,q} = R^p \Gamma_Y(R^q f_* \mathcal{F}) = H^p(Y, R^q f_* \mathcal{F}). \qquad \square$$

If the spectral sequence degenerates at E_2, and e.g., the groups are vector spaces, so that the limiting groups may be (noncanonically) identified with their graded modules, then

$$H^n(X, \mathcal{F}) \simeq \bigoplus_{p+q=n} H^p(Y, R^q f_* \mathcal{F}).$$

This spectral sequence can also be directly derived from the hyper-cohomology spectral sequence. This just amounts to explicitly tracking some steps of the proof of the Grothendieck spectral sequence, but still it may be a useful exercise. Given a sheaf $\mathcal{F}$ on X, let $0 \to \mathcal{F} \to \mathcal{I}^\bullet$ be an injective resolution. As we have already noted, the sheaves $f_* \mathcal{I}^\bullet$ are flabby. We consider now the hypercohomology spectral sequence of Theorem 5.25 for the global section functor $\Gamma_Y \colon \mathfrak{Sh}_Y \to \mathfrak{Ab}$, as applied to the complex $f_* \mathcal{I}^\bullet$. Since the sheaves $f_* \mathcal{I}^\bullet$ are flabby, hence Γ_Y-acyclic, by Proposition 5.20 we can compute the hypercohomology of the complex $f_* \mathcal{I}^\bullet$ directly on it, without replacing it with a quasi-isomorphic complex of injectives, so that the spectral sequence converges to

$$\mathbb{R}^i \Gamma_Y(f_* \mathcal{I}^\bullet) \simeq H^i(\Gamma_Y(f_* \mathcal{I}^\bullet)) \simeq H^i(\Gamma_X(\mathcal{I}^\bullet)) = H^i(X, \mathcal{F}).$$

Concerning the first pages of the two spectral sequences, from equation (5.17) we see that

$${'E}_2^{p,q} \simeq R^p \Gamma_Y(H^q(f_* \mathcal{I}^\bullet)) = H^p(Y, R^q f_* \mathcal{F}),$$

i.e., we get the second page of the Leray spectral sequence. On the other hand, from equation (5.14) we obtain

$$E_1^{p,q} = H^q(Y, f_* \mathcal{I}^p),$$

and as the sheaves $f_* \mathcal{I}^\bullet$ are flabby, we have

$$E_1^{p,0} = H^0(Y, f_* \mathcal{I}^p) \simeq \Gamma_X(\mathcal{I}^p), \quad E_1^{p,q} = 0 \text{ for } q > 0.$$

As a result, this spectral sequence degenerates at E_2, and consistently,

$$E_2^{p,0} = H^p(\Gamma_X(\mathcal{I}^\bullet)) = H^p(X, \mathcal{F}), \quad E_2^{p,q} = 0 \text{ for } q > 0.$$

5.6.6. The Hochschild–Serre spectral sequence

If $\mathfrak{g}$ is a Lie algebra and $\mathfrak{h}$ is an ideal in $\mathfrak{g}$, the quotient $\mathfrak{q} = \mathfrak{g}/\mathfrak{h}$ has a natural structure of Lie algebra. Let M be a representation of $\mathfrak{g}$. All these data give rise to a spectral sequence converging to the Lie algebra cohomology $H^\bullet(\mathfrak{g}, M)$ [32].

Theorem 5.38. *Let $\mathfrak{g}$ be a Lie algebra over a commutative ring R; moreover, let M be a representation of $\mathfrak{g}$, and $\mathfrak{h}$ an ideal of $\mathfrak{g}$. Then, there exists a spectral sequence converging to $H^\bullet(\mathfrak{g}, M)$, whose second page is*

$$E_2^{p,q} = H^p(\mathfrak{q}, H^q(\mathfrak{h}, M)).$$

Proof. Let us consider the exact sequence

$$0 \to \mathfrak{h} \to \mathfrak{g} \to \mathfrak{q} \to 0$$

of Lie algebras. First we note that the categories $\mathrm{Rep}(\mathfrak{g})$ and $\mathrm{Rep}(\mathfrak{q})$ are related by two functors. Given a representation N of $\mathfrak{q}$, the composition

$$\mathfrak{g} \to \mathfrak{q} \to \mathrm{End}_R(N)$$

defines a representation of $\mathfrak{g}$, so that there is a functor $G \colon \mathrm{Rep}(\mathfrak{q}) \to \mathrm{Rep}(\mathfrak{g})$, which turns out be exact. On the other hand, in the same way a representation M of $\mathfrak{g}$ can be regarded as a representation of $\mathfrak{h}$, and it is not difficult to check that the invariant submodule $M^{\mathfrak{h}}$ is a representation of $\mathfrak{q}$. Therefore we have a functor

$$(-)^{\mathfrak{h}} \colon \mathrm{Rep}(\mathfrak{g}) \to \mathrm{Rep}(\mathfrak{q}).$$

One can check by some direct computations that G is left adjoint to $(-)^{\mathfrak{h}}$, so that the latter functor sends injectives objects of $\mathrm{Rep}(\mathfrak{g})$ into injective objects of $\mathrm{Rep}(\mathfrak{q})$ (see Exercise 9 in Chapter 2).

So we consider the composition of functors

$$\mathrm{Rep}(\mathfrak{g}) \xrightarrow{(-)^{\mathfrak{h}}} \mathrm{Rep}(\mathfrak{q}) \xrightarrow{(-)^{\mathfrak{q}}} R\text{-mod}.$$

Since $(-)^{\mathfrak{q}} \circ (-)^{\mathfrak{h}} = (-)^{\mathfrak{g}}$, the hypotheses of Grothendieck's Theorem 5.28 are satisfied, and one has a spectral sequence converging to $R^\bullet M^{\mathfrak{g}} = H^\bullet(\mathfrak{g}, M)$, whose second page is[e]

$$E_2^{p,q} = R^p(R^q M^{\mathfrak{h}})^{\mathfrak{q}} = H^p(\mathfrak{q}, H^q(\mathfrak{h}, M)). \qquad \square$$

---

[e]If we regard $(-)^{\mathfrak{h}}$ as a functor $\mathrm{Rep}(\mathfrak{g}) \to \mathrm{Rep}(\mathfrak{q})$, its derived functors $H^q(\mathfrak{h}, -)$ take values in $\mathrm{Rep}(\mathfrak{q})$ as well, so that the cohomology modules $H^q(\mathfrak{h}, M)$ are representations

Remark 5.39. It can be shown that the above spectral sequence is the same as that induced by the so-called Hochschild–Serre filtration of the Chevalley–Eilenberg complex, defined as

$$F_p C^q(\mathfrak{g}, M) = \{\xi \in C^q(\mathfrak{g}, M) \text{ which vanish when at least } q - p + 1$$

of their arguments are in $\mathfrak{h}\}$.

This spectral sequence was originally introduced in [32] in this way.

5.6.7. The Frölicher spectral sequence

Another application of spectral sequences is related to the Hodge decomposition of the de Rham cohomology of a compact Kähler manifold. Here we only give a cursory treatment; for a more complete discussion the reader may refer, e.g., to [62]. We recall that a Kähler manifold is a pair (X, g), where X is a complex manifold, and g is an Hermitian metric on X, such that the $(1,1)$-form ω associated to g is closed, $d\omega = 0$ [38, 62].[f] The Dolbeault cohomology groups $H_{\bar{\partial}}^{p,q}(X)$ were defined in Section 4.4. The celebrated Hodge Decomposition Theorem states that for a compact Kähler manifold one has an isomorphism

$$H_{dR}^k(X, \mathbb{C}) \simeq \bigoplus_{p+q=k} H_{\bar{\partial}}^{p,q}(X)$$

for $k = 1, \ldots, 2 \dim_{\mathbb{C}} X$ (see, for instance, [62, Proposition 6.11]). The subspaces $H_{\bar{\partial}}^{p,q}(X)$ satisfy the Hodge symmetry

$$H_{\bar{\partial}}^{p,q}(X) = \overline{H_{\bar{\partial}}^{q,p}(X)}.$$

---

of q. When $\mathfrak{g}$ is free over R, this can also be checked directly by realizing $H^\bullet(\mathfrak{h}, M)$ as the cohomology groups of the Chevalley–Eilenberg complex, see Example 2.31. One gives a q-module structure to the modules of p-cochains by

$$(x\xi)(y_1, \ldots, y_p) = \sum_{i=1}^p (-1)^{i-1} \xi(y_1, \ldots, [\tilde{x}, y_i], \ldots, y_p),$$

where $x \in \mathfrak{q}$, $\xi \in \mathrm{Hom}_R(\wedge^p\mathfrak{h}, M)$, and $y_i \in \mathfrak{h}$; $\tilde{x}$ is an element in $\mathfrak{g}$ which projects to x. This passes to cohomology.

[f] In local homomorphic coordinates $(z^1, \ldots, z^n)$, if

$$g = \sum_{j,k=1}^n g_{jk} \, dz^j \otimes d\bar{z}^k,$$

then

$$\omega = i \sum_{j,k=1}^n g_{jk} \, dz^j \wedge d\bar{z}^k$$

(i is the imaginary unit). The condition $d\omega = 0$ is equivalent to the fact that the complex structure of X is parallelly transported by the Levi-Civita connection of g.

By means of the Hodge decomposition, one defines the Hodge filtration of the complex de Rham cohomology

$$F^p H^k_{\mathrm{dR}}(X, \mathbb{C}) = \bigoplus_{r \geq p} H^{r,k-r}_{\bar{\partial}}(X). \tag{5.28}$$

Let A^k_X be the sheaves of complex valued C^∞ differential forms on X, i.e.,

$$A^k_X = \bigoplus_{p+q=k} \Omega^{p,q}_X;$$

these sheaves are filtered by

$$F^p A^k_X = \bigoplus_{r \geq p} \Omega^{r,k-r}_X, \tag{5.29}$$

and this induces a filtration of the spaces of global sections $A^k_X(X)$.

We shall need the following result [62, Proposition 7.5], which expresses the compatibility of the filtration of the de Rham complex with that of the de Rham cohomology groups.

Proposition 5.40. *For every p and k one has*

$$F^p H^k_{\mathrm{dR}}(X, \mathbb{C}) = \frac{\ker d \colon F^p A^k_X(X) \to F^p A^{k+1}_X(X)}{\operatorname{im} d \colon F^p A^{k-1}_X(X) \to F^p A^k_X(X)}.$$

Definition 5.41. The Frölicher spectral sequence is the spectral sequence associated with the filtration (5.29) of the de Rham complex.

Theorem 5.42. *The Frölicher spectral sequence of a compact Kähler manifold degenerates at the first page.*

Proof. We compute the pages E_∞ and E_1 of the Frölicher spectral sequence. From equation (5.7) we see that

$$E^{p,q}_0 = \Omega^{p,q}_X(X);$$

the differential $d_0 \colon E^{p,q}_0 \to E^{p,q+1}_0$ is $\bar{\partial}$, so that

$$E^{p,q}_1 = H^{p,q}_{\bar{\partial}}(X). \tag{5.30}$$

On the other hand, since the spectral sequence converges to the complexified de Rham cohomology of X,

$$E^{p,q}_\infty = F^p H^{p+q}_{\mathrm{dR}}(X, \mathbb{C}) / F^{p+1} H^{p+q}_{\mathrm{dR}}(X, \mathbb{C})$$

(cf. equation (5.8)), where this filtration is induced by the filtration (5.29). By Proposition 5.40 these quotients are the same as those of the filtration (5.28), so that

$$E^{p,q}_\infty \simeq H^{p,q}_{\bar{\partial}}(X).$$

Thus,

$$E_\infty^{p,q} \simeq E_1^{p,q}$$

for all p and q, and in particular, $\dim E_\infty^{p,q} = \dim E_1^{p,q}$. This implies that all differentials d_i of the spectral sequence vanish for $i \geq 1$. $\square$

Remark 5.43. The Frölicher spectral sequence makes perfect sense also for a complex manifold without a Kähler structure, but in general it does not degenerate at the first page. This fact is actually equivalent to equation (5.30), and is weaker than the Hodge Decomposition Theorem. See [62] for a discussion.

5.6.8. Künneth spectral sequences

This section will use the notions of *flat module* and *projective resolution* (see Section 2.7). Moreover, we shall consider *homology spectral sequences*; this is just a formal difference with respect to the cohomology spectral sequences we have consider so far — the differentials *decrease* the degree by one instead of increasing it. The bidegree of the differential acting on the rth page is given by

$$d_r \colon E_{p,q}^r \to E_{p-r,q+r-1}^r.$$

In its simplest form, the gist of the Künneth spectral sequence, given a homology complex $F_\bullet$ and a module M, is to compare the homology of $F_\bullet$ with that of $F_\bullet \otimes M$.

Let $F_\bullet$ be a bounded above complex of flat R-modules, where R is a commutative ring with unity, i.e., each F_n is a flat R-module, and there is $n_0 \in \mathbb{Z}$ such that $F_n = 0$ for $n < n_0$.[g] Let M be an R-module.

Theorem 5.44. *There is a homology spectral sequence converging to*

$$H_\bullet(F_\bullet \otimes_R M)$$

whose second page is

$${}'E_{p,q}^2 = \operatorname{Tor}_p^R(H_q(F_\bullet), M).$$

---

[g]So $F_\bullet$ is a complex of the form

$$\cdots \to F_{m+1} \to F_m \to \cdots \to F_{n_0+1} \to F_{n_0} \to 0.$$

Proof. Let $Q_\bullet$ be a projective resolution of M, and consider the double complex $K_{\bullet,\bullet} = F_\bullet \otimes_R Q_\bullet$. Since the modules $F_\bullet$ are flat, we have

$$E^1_{p,q} = H_q(F_p \otimes_R Q_\bullet) \simeq F_p \otimes_R H_q(Q_\bullet) = \begin{cases} F_p \otimes_R M & \text{if } q = 0, \\ 0 & \text{if } q > 0; \end{cases}$$

as a result, the first spectral sequence degenerates at the second page:

$$E^2_{p,q} = \begin{cases} H_p(F_\bullet \otimes_R M) & \text{if } q = 0, \\ 0 & \text{if } q > 0, \end{cases}$$

and converges to $H_\bullet(F_\bullet \otimes_R M)$.

On the other hand, as projective modules are flat, we also have

$$H_q(F_\bullet \otimes_R Q_p) \simeq H_q(F_\bullet) \otimes_R Q_p.$$

Therefore, the second term of the second spectral sequence is

$$'E^2_{p,q} = H_p(H_q(F_\bullet) \otimes_R Q_\bullet) = \mathrm{Tor}^R_p(H_q(F_\bullet), M)$$

since $Q_\bullet$ is a projective resolution of M. $\qquad\square$

An immediate application of this spectral sequence is the exact sequence known as *Universal Coefficient Theorem*. For every n, let B_n and Z_n be the *n-boundaries* and *n-cycles* of the complex $F_\bullet$, i.e.,

$$B_n = \mathrm{im}(d_{n+1} \colon F_{n+1} \to F_n),$$
$$Z_n = \ker(d_n \colon F_n \to F_{n-1}).$$

Corollary 5.45. *In addition to the previous assumptions, suppose that the boundary modules B_n are flat R-modules. Then, for every n, there is an exact sequence*

$$0 \to H_n(F_\bullet) \otimes_R M \to H_n(F_\bullet \otimes_R M) \to \mathrm{Tor}^R_1(H_{n-1}(F_\bullet), M) \to 0. \quad (5.31)$$

Proof. Since the boundary modules B_n are flat, the cycle modules Z_n are flat as well.[h] Then, the exact sequence

$$0 \to B_n \to Z_n \to H_n(F_\bullet) \to 0$$

---

[h]Indeed one has an exact sequence

$$0 \to Z_n \to F_n \to B_{n-1} \to 0$$

where F_n and B_{n-1} are flat, so that Z_n is flat as well (see, e.g., [41, Proposition XVI.3.4]).

provides a flat resolution of the R-modules $H_n(F_\bullet)$ — i.e., these R-modules have homological dimension 1. As a result, the only nonvanishing columns of the second page $'E_2$ of the second spectral sequence are those corresponding to $p = 0$ and $p = 1$. Now the homology analogue of the exact sequence in Exercise 1 in this chapter reads

$$0 \to {}''E^2_{0,n} \to H_n(K) \to {}''E^2_{1,n-1} \to 0$$

which produces the exact sequence in the claim, keeping in mind that $\mathrm{Tor}^R_0(N, M) \simeq N \otimes_R M$ for all R-modules M and N. $\qquad\square$

Remark 5.46. The exact sequence (5.31), and the companion exact sequence (5.33), split, albeit nonnaturally. For a discussion of this issue see, e.g., [64].

Example 5.47. Let us see how Corollary 5.45 produces the Universal Coefficient Theorem in singular homology with coefficients in a principal ideal domain P. In that case, $F_\bullet$ is the complex $C_\bullet(X, P)$ of singular chains of a topological space X with coefficients in P [29, 57], i.e.,

$$H_n(X, P) = H_n(C_\bullet(X, P)).$$

By definition, every P-module $C_n(X, P)$ is free, hence flat; and since any submodule of a free module over a principal ideal domain is free,[i] the boundary modules are free, hence flat, as well. Thus, all the hypotheses of Corollary 5.45 are met, and for every P-module M and every n one has the exact sequence

$$0 \to H_n(X, P) \otimes_P M \to H_n(X, M) \to \mathrm{Tor}^P_1(H_{n-1}(X, P), M) \to 0.$$

Thus, when the ring R is a principal ideal domain, the R-modules $H_n(C_\bullet(X, R))$ have homological dimension 1, so that the spectral sequence of Theorem 5.44 is replaced by the previous exact sequence.

Theorem 5.44 and Corollary 5.45 have a kind of "dual" counterparts for cohomology, in the sense of Theorem 5.48 below. Let $Q_\bullet$ be a homology complex of projective R-modules, M an R-module and $I^\bullet$ an injective

---

[i]This fact is somehow nontrivial to prove; see, e.g., [30, Theorem I.5.1].

resolution of M. Let us introduce the double complex

$$K^{p,q} = \mathrm{Hom}_R(Q_p, I^q)$$

with the obvious differential (note that for fixed q, $\mathrm{Hom}_R(Q_\bullet, I^q)$ is a cohomology complex). Since the functors $\mathrm{Hom}_R(-, I^\bullet)$ are exact, we have

$$H^q(\mathrm{Hom}_R(Q_p, I^\bullet)) \simeq \mathrm{Hom}(Q_p, H^q(I^\bullet)) = \begin{cases} \mathrm{Hom}(Q_p, M) & \text{if } q = 0, \\ 0 & \text{if } q > 0; \end{cases}$$

then the first spectral sequence of the double complex $K^{\bullet\bullet}$ degenerates at the second page, and converges to $H^n(K) \simeq H^n(\mathrm{Hom}_R(Q_\bullet, M))$. The second page of the second spectral sequence reads

$$'E_2^{p,q} = H^p(\mathrm{Hom}(H_q(Q_\bullet), I^\bullet)) = \mathrm{Ext}^p(H_q(Q_\bullet), M) \qquad (5.32)$$

so that we have the following theorem.

Theorem 5.48. *There is a spectral sequence, whose second page is given in equation (5.32), which converges to $H^\bullet(\mathrm{Hom}_R(Q_\bullet, M))$.*

As a corollary, we have a dual version of the Universal Coefficient Theorem.

Corollary 5.49. *In addition to the previous hypotheses, assume that the boundary modules B_n of $Q^\bullet$ are projective R-modules. Then, for every n, there is an exact sequence*

$$0 \to \mathrm{Ext}^1(H_{n-1}(Q_\bullet), M) \to H^n(\mathrm{Hom}_R(Q_\bullet, M))$$
$$\to \mathrm{Hom}_R(H_n(Q_\bullet), M) \to 0. \qquad (5.33)$$

Proof. Since the boundary modules are projective, the cycle modules Z_n are projective as well,[j] and therefore the exact sequences

$$0 \to B_n \to Z_n \to H_n(Q_\bullet) \to 0$$

provide projective resolutions of the R-modules $H_n(Q_\bullet)$. As a result, the spectral sequence of Theorem 5.48 only has the zeroth and first columns. Then, the exact sequence in Exercise 1 in this chapter produces the exact sequence in the claim. $\qquad \square$

---

[j]Indeed since every module B_{n-1} is projective, every exact sequence

$$0 \to Z_n \to Q_n \to B_{n-1} \to 0$$

splits; and as every Q_n is a summand of a free module, the same is true for Z_n, which therefore is projective.

Example 5.50. The "dual" to Example 5.47 establishes the relation between singular homology and cohomology. Take for $Q_\bullet$ the singular chain complex $C_\bullet(X, P)$, where X is a topological space and P is a principal ideal domain,[k] and let M be a P-module. By definition, the groups $H^n(\text{Hom}_P(C_\bullet(X, P), M))$ are the singular cohomology groups of X with coefficients in M [29, 57]. Again, the boundary modules are free, and hence projective. Then, the exact sequence (5.33) gives

$$0 \to \text{Ext}^1_P(H_{n-1}(X, P), M) \to H^n(X, M)$$

$$\to \text{Hom}_P(H_n(X, P), M) \to 0.$$

And again, this exact sequence splits, albeit nonnaturally.

The results we have seen so far in this section can be further generalized considering tensor products of complexes. We give here some results without proofs, which anyway can be given along the lines of what we saw in this Section (for a full treatment see, e.g., [54]). For simplicity we assume that all complexes are positive.

Theorem 5.51. *Let $S_\bullet$ and $T_\bullet$ be homology complexes of R-modules, and assume that all modules S_n are flat. Then, there is a homology spectral sequence converging to the cohomology of the total complex of the tensor product $S_\bullet \otimes_R T_\bullet$ whose second page is*

$$E^2_{p,q} = \bigoplus_{m+n=q} \text{Tor}^R_p(H_m(S_\bullet), H_n(T_\bullet)).$$

Corollary 5.52 (Generalized Universal Coefficient Theorem). *Assume additionally that the boundary and cycle modules of $S_\bullet$ are all flat. Then, for every n, there is an exact sequence*

$$0 \to \bigoplus_{p+q=n} H_p(S_\bullet) \otimes_R H_q(T_\bullet) \to H_n(S_\bullet \otimes_R T_\bullet)$$

$$\to \bigoplus_{p+q=n-1} \text{Tor}^R_1(H_p(S_\bullet), H_q(T_\bullet)) \to 0.$$

Corollary 5.53. *Let $S_\bullet$ and $T_\bullet$ be homology complexes of R-modules, and assume that all modules $Z_n(S_\bullet)$ and $H_n(S_\bullet)$ are projective. Then, for every n, there is an isomorphism*

$$\bigoplus_{p+q=n} H_p(S_\bullet) \otimes_R H_q(T_\bullet) \simeq H_n(S_\bullet \otimes_R T_\bullet).$$

---

[k]Actually, it is enough to assume that the ring R is *hereditary*, i.e., every submodule of a projective R-module is projective.

Example 5.54. Let X, Y be topological spaces, and let P be a principal ideal domain. Then, Corollary 5.52 yields the Künneth Theorem for the singular homology of the cartesian product $X \times Y$ with coefficients in P:

$$0 \to \bigoplus_{p+q=n} H_p(X,P) \otimes_P H_q(Y,P) \to H_n(X \times Y, P)$$

$$\to \bigoplus_{p+q=n-1} \mathrm{Tor}_1^P(H_p(X,P), H_q(Y,P)) \to 0.$$

This is obtained by taking $S_\bullet = C_\bullet(X,P)$ and $T_\bullet = C_\bullet(Y,P)$. One also needs to identify $H_n(C_\bullet(X,P) \otimes_R C_\bullet(Y,P))$ with the homology of $X \times Y$ — see [29] for details.

If the Tor groups vanish (for instance, when P is a field, or more generally when the homology groups of one of the two spaces are all flat P-modules) one obtains the topological Künneth theorem in its simplest form:

$$H_n(X \times Y, P) \simeq \bigoplus_{p+q=n} H_p(X,P) \otimes_P H_q(Y,P).$$

Example 5.55. With the material in this section (adapted to the case of cohomology) and in Chapter 4, one can prove a Künneth Theorem for the sheaf cohomology of quasi-coherent sheaves on the product of two schemes. Let X and Y be compact and separated[1] schemes over a field $\Bbbk$, and let $\mathcal{F}$ and $\mathcal{G}$ be quasi-coherent sheaves on X and Y, respectively. Denote by π_1, π_2 the projections onto the factors of the product $X \times_\Bbbk Y$. For every $n \geq 0$ there are isomorphisms

$$H^n(X \times_\Bbbk Y, \pi_1^* \mathcal{F} \otimes_{\mathcal{O}_{X \times_\Bbbk Y}} \pi_2^* \mathcal{G}) \simeq \bigoplus_{p+q=n} H^p(X, \mathcal{F}) \otimes_\Bbbk H^q(Y, \mathcal{G}).$$

For a discussion and a proof of this result the reader can refer, e.g., to [49, Lemmas 32.29.1 and 32.29.2].

5.7. Additional Exercises

1. Let K be a filtered complex such that the only nonzero columns of the second page of the corresponding first-quadrant spectral sequence are

---

[1]It is actually enough that the diagonal morphisms of both X and Y are affine; a scheme morphism $f: X \to Z$ is said to be affine if Z has an open affine cover $\{U_i\}$ such that $f^{-1}(U_i)$ is an affine subscheme of X for every i, see, e.g., [28, Exercise II.5.17].

the first two, i.e., $E_2^{p,q} = 0$ for $p \neq 0, 1$. Prove that for each n there is an exact sequence

$$0 \to E_2^{1,n-1} \to H^n(K) \to E_2^{0,n} \to 0.$$

Hint: use equation (5.8).

2. Let $\mathfrak{A} \xrightarrow{F} \mathfrak{B} \xrightarrow{G} \mathfrak{C}$ be left exact functors between abelian categories, and assume that both $\mathfrak{A}$ and $\mathfrak{B}$ have enough injectives.

   (a) Prove that if G is exact, then $R^q(G \circ F) \simeq G \circ R^q F$ for all $q \geq 0$.
   (b) Prove that if F is exact and send injective objects to G-acyclic objects, then $R^p(G \circ F) \simeq R^p G \circ F$ for all $p \geq 0$.

3. Let $f : R \to S$ be a morphism of commutative rings, and take an R-module P and a flat S-module M. Use Exercise 13 in Chapter 2 to construct a spectral sequence whose second page is

$$E_2^{p,q} = \text{Ext}_R^p(P, \text{Ext}_S^q(M, N)).$$

4. Provide a proof of Theorem 5.29, suitably modifying the proof of Theorem 5.28.

5. Here X, Y, F will be paracompact, locally Euclidean topological spaces. Let $\pi : X \to Y$ be a locally trivial fibre bundle on Y with standard fibre F, i.e., Y has an open cover $\{U_j\}$ such that every $\pi^{-1}(U_j)$ is homeomorphic to $U_j \times F$. On any of the spaces X, Y, F denote by $\mathbb{Z}$ be the constant sheaf with stalk the integers.

   (a) In view of Proposition 4.34, use a homotopy argument to show that $H^i(\pi^{-1}(U_j), \mathbb{Z}) \simeq H^i(F, \mathbb{Z})$ for all i.
   (b) Assume now that F is contractible, and use the description of the higher direct images starting at p. 89 to show that $R^i \pi_* \mathbb{Z} = 0$ for $i > 0$.
   (c) Use the Leray spectral sequence to show that $H^i(X, \mathbb{Z}) \simeq H^i(Y, \mathbb{Z})$ for $i \geq 0$.

6. (a) With reference to Exercise 8 of Chapter 4, and using Proposition 4.25, compute the higher direct images $R^i \pi_* \mathbb{R}$, where $\mathbb{R}$ is the constant sheaf.
   (b) Compute all terms of the second page of the corresponding Leray spectral sequence and show that $d_2 = 0$.
   (c) Use the Leray spectral sequence to compute the cohomology of $\mathbb{F}_n$ with coefficients in $\mathbb{R}$.

7. In analogy with Example 5.11, provide a proof of the Dolbeault Theorem 4.44 using spectral sequences.

8. $U(n)$ is the group of unitary $n \times n$ complex matrices. For every $n \geq 2$ there is an embedding $U(n-1) \to U(n)$, and $U(n)/U(n-1) \simeq S^{2n-1}$. Use induction on n and the Leray spectral sequence to compute the groups $H^i(U(n), \mathbb{Z})$.

Hint: compute first the cohomology of $U(2)$; then guess a formula for the cohomology of $U(n-1)$, and use induction.

9. Let X be a smooth, irreducible projective curve X over a field (for instance, we can work with the field $\mathbb{C}$ and look at X as a compact connected complex manifold of dimension 1). Fix a point p and consider the exact sequence

$$0 \to \mathcal{I}_p \to \mathcal{O}_X \to \mathcal{O}_p \to 0,$$

where $\mathcal{I}_p$ is the ideal sheaf of p (sheaf of sections of $\mathcal{O}_X$ than vanish at p) and the quotient sheaf $\mathcal{O}_p$ is the structure sheaf of the point. Note that $\mathcal{H}om_{\mathcal{O}_X}(\mathcal{O}_p, \mathcal{O}_X) = 0$ (the dual of a torsion module is 0).

(a) Prove that the sheaf $\mathcal{E}xt^1_{\mathcal{O}_X}(\mathcal{O}_p, \mathcal{O}_X)$ is supported on p, and that $\mathcal{E}xt^1_{\mathcal{O}_X}(\mathcal{I}_p, \mathcal{O}_X) = 0$.

(b) Compute the second page of the local-to-global spectral sequence for the local Ext sheaves $\mathcal{E}xt^p_{\mathcal{O}_X}(\mathcal{O}_p, \mathcal{O}_X)$ and check that $d_2 = 0$.

(c) Use the local-to-global spectral sequence to compute $\mathrm{Ext}^1_{\mathcal{O}_X}(\mathcal{O}_p, \mathcal{O}_X)$.

Hint: you will have to use the fact that the Ext sheaves of two $\mathcal{O}_X$-modules over a noetherian scheme of dimension n vanish in degree higher that n. This is a purely algebraic fact, see [45, Section 18]; see also [28, Exercise III.6.5].

10. Let X be a complex manifold.

(a) Prove that the holomorphic de Rham sheaf complex $(\Omega^\bullet_X, \partial)$ is quasi-isomorphic to the complexified de Rham complex $(A^\bullet_X, d)$.
    Hint: remember that for every p, $\Omega^{p,\bullet}_X$ is a resolution of Ω^p_X (equation (4.17)).

(b) Deduce that the holomorphic de Rham sheaf complex is a resolution of the constant sheaf $\mathbb{C}$.

(c) Prove that $\mathbb{H}^k(X, \Omega^\bullet_X) \simeq H^k(X, \mathbb{C})$ for every k.

11. Let X be an irreducible topological space, and let $\mathbb{Z}$, $\mathbb{Z}_2$ be the constant sheaves with stalks the corresponding groups.

(a) Use the local-to-global spectral sequence to compute $\mathrm{Ext}^1_{\mathbb{Z}}(\mathbb{Z}_2, \mathbb{Z}_2)$.

(b) Use this calculation to provide an example of a flabby sheaf of abelian groups which is not injective.

12. Let $X = \mathbb{RP}^2 \times \mathbb{RP}^2$, where $\mathbb{RP}^2$ is the real projective plane

$$\frac{\mathbb{R}^2 - \{0\}}{\mathbb{R}^*}$$

with the induced the Euclidean quotient topology.

(a) Use the Künneth Theorem for singular homology (Example 5.54) to compute the homology of X with coefficients in $\mathbb{Z}$.

(b) Use the Universal Coefficient Theorem (Example 5.50) to compute the cohomology of X with coefficients in $\mathbb{Z}_2$.

Hint: $\mathbb{RP}^2$ may be regarded as the quotient $S^2/\mathbb{Z}_2$, where $\mathbb{Z}_2$ acts by identifying antipodal points. Also, you will need Exercise 15 of Chapter 2.

13. A morphism $\phi \colon \mathfrak{h} \to \mathfrak{g}$ of Lie algebras over a ring R is a morphism of R-modules which is compatible with the two brackets. Note that a representation of $\mathfrak{g}$ induces via ϕ a representation of $\mathfrak{h}$. Show that, given a representation M of $\mathfrak{g}$, ϕ induces a morphism

$$\phi^\bullet \colon H^\bullet(\mathfrak{g}, M) \to H^\bullet(\mathfrak{h}, M)$$

of δ-functors.

14. Write the five-term sequence for the Hochschild–Serre spectral sequence associated with an ideal $\mathfrak{k}$ in a Lie algebra $\mathfrak{g}$.

15. (a) Prove that the space $\mathfrak{gl}(n)$ of $n \times n$ matrices with entries in a field $\Bbbk$, equipped with the bracket given by the commutator of matrices, is a Lie algebra over $\Bbbk$.

(b) Prove that the subspace $\mathfrak{sl}(n)$ formed by the matrices with vanishing trace is an ideal in $\mathfrak{gl}(n)$, and compute the corresponding quotient $\mathfrak{q}$.

(c) Specialize the five-term sequence of Exercise 15 to this case.

(d) Compute the group $H^2(\mathfrak{q}, \mathfrak{sl}(n))$.

16. Let π_1, π_2 be the projections onto the factors of $\mathbb{P}^1_\Bbbk \times_\Bbbk \mathbb{P}^1_\Bbbk$, where $\mathbb{P}^1_\Bbbk$ is the projective line over a field $\Bbbk$. Define the sheaves

$$\mathcal{O}_{\mathbb{P}^1_\Bbbk \times_\Bbbk \mathbb{P}^1_\Bbbk}(p, q) = \pi_1^* \mathcal{O}_{\mathbb{P}^1_\Bbbk}(p) \otimes_{\mathcal{O}_{\mathbb{P}^1_\Bbbk \times_\Bbbk \mathbb{P}^1_\Bbbk}} \pi_2^* \mathcal{O}_{\mathbb{P}^1_\Bbbk}(q).$$

Use the Künneth Theorem in Example 5.55 to compute the groups

$$H^n(\mathbb{P}^1_\Bbbk \times_\Bbbk \mathbb{P}^1_\Bbbk, \mathcal{O}_{\mathbb{P}^1_\Bbbk \times_\Bbbk \mathbb{P}^1_\Bbbk}(p, q)).$$

Chapter 6

Epilogue: Derived Categories

The natural development of the theory of derived functors is the notion of *derived category,* due originally to Alexander Grothendieck and his student Jean-Luis Verdier. A first summary of the theory was presented by Verdier in 1977 in the SGA $4\frac{1}{2}$ volume [60], and a fuller treatment was given in his thesis, published only in 1996 [61]. The notion of derived category simplifies very much the treatment of derived functors; for instance, relations between derived functors that are described by spectral sequences often in derived category become isomorphisms (we shall give an example below, see Theorem 6.5).

A rather self-contained introduction to derived categories may be found in [4, Appendix A]. A good introduction is also the first chapter of [36]. Here, we shall only try to give a basic idea of what a derived category is, and how a derived functor between derived categories is defined.

Derived categories. Given an abelian category $\mathfrak{A}$, let $K(\mathfrak{A})$ be the category of complexes of objects in $\mathfrak{A}$; it is an abelian category. The derived category $D(\mathfrak{A})$ will have the same objects, i.e., complexes of objects in $\mathfrak{A}$, but the morphisms are defined in a subtler way, basically first modding by homotopies, and then localizing with respect to quasi-isomorphisms, in a sense that we shall make precise later on.

We shall denote by $\operatorname{Hom}_{\mathfrak{A}}^0(a^\bullet, b^\bullet)$ the subgroup of morphisms in $K(\mathfrak{A})$, $f \colon a^\bullet \to b^\bullet$, that are homotopic to the zero morphism (see Section 1.2.3 for the notion of homotopy between morphisms of complexes).

Definition 6.1. The homotopy category of $\mathfrak{A}$ is the category $\mathfrak{Ht}(\mathfrak{A})$ whose objects are the complexes of objects in $\mathfrak{A}$, and the groups of morphisms are the quotient groups

$$\mathrm{Hom}_{\mathfrak{Ht}(\mathfrak{A})}(a^\bullet, b^\bullet) = \mathrm{Hom}_{\mathfrak{A}}(a^\bullet, b^\bullet)/\mathrm{Hom}_{\mathfrak{A}}^0(a^\bullet, b^\bullet).$$

Morphisms in the derived category, for instance a morphism from $a^\bullet \to c^\bullet$, will be equivalence classes of diagrams in $K(\mathfrak{A})$

$$\begin{array}{ccc} & b^\bullet & \\ {}^{\phi}\swarrow & & \searrow \\ a^\bullet & & c^\bullet \end{array} \qquad (6.1)$$

where ϕ is a quasi-isomorphism (see Lemma 5.16 for this notion) — i.e., morphisms are morphisms of complexes up to quasi-isomorphisms. This diagram and another diagram "from $a^\bullet$ to $c^\bullet$", say $a^\bullet \xleftarrow{\psi} \bar{b}^\bullet \to c^\bullet$, are considered to be equivalent if there is a diagram

$$\begin{array}{ccccc} & & e^\bullet & & \\ & {}^{\chi}\swarrow & & \searrow^{\lambda} & \\ & b^\bullet & & \bar{b}^\bullet & \\ {}^{\phi}\swarrow & & \times & & \searrow \\ a^\bullet & & {}_{\psi} & & c^\bullet \end{array}$$

where χ and λ are quasi-isomorphisms, which commutes in $\mathfrak{Ht}(\mathfrak{A})$, i.e., commutes up to homotopies.

Definition 6.2. $D(\mathfrak{A})$, the derived category of $\mathfrak{A}$, is the category whose objects are complexes in $\mathfrak{A}$, while the morphisms are classes of diagrams (6.1) that are equivalent in the sense just described.

The composition of two morphisms $a^\bullet \leftarrow b^\bullet \to c^\bullet$ and $c^\bullet \leftarrow \bar{b}^\bullet \to \bar{c}^\bullet$ is the morphism $a^\bullet \leftarrow p^\bullet \to \bar{c}^\bullet$ given by the diagram

$$\begin{array}{ccccc} & & p^\bullet & & \\ & {}^{\pi}\swarrow & & \searrow & \\ & b^\bullet & & \bar{b}^\bullet & \\ {}^{\phi}\swarrow & & \searrow \quad {}^{\rho}\swarrow & & \searrow \\ a^\bullet & & c^\bullet & & \bar{c}^\bullet \end{array}$$

where $p^\bullet$ is the pullback (fibred product)[a] of $b^\bullet$ and $\bar{b}^\bullet$ over $c^\bullet$ in the category $\mathfrak{Ht}(\mathfrak{A})$ (the complex $p^\bullet$ is described explicitly in [4, Lemma A.29]). The morphism π is a quasi-isomorphism because ρ is a quasi-isomorphism, and therefore the composition $\phi \circ \pi$ is a quasi-isomorphism as well.

Remark 6.3. The derived category $D(\mathfrak{A})$ is additive but not abelian; the existence of kernels and cokernels is replaced by a structure — the cone of a morphism — that makes it into a *triangulated category* [4].

Derived functors. Let $K^+(\mathfrak{A})$ be the full subcategory of $K(\mathfrak{A})$ made by *left bounded complexes*, i.e., complexes $a^\bullet$ for which there is $n_0 \in \mathbb{Z}$ such that $a^n = 0$ for $n < n_0$. Then one can also define categories $\mathfrak{Ht}^+(\mathfrak{A})$ and $D^+(\mathfrak{A})$ whose objects are complexes in $K^+(\mathfrak{A})$. As a preparation to the introduction of the notion of right derived functor between derived categories, we note that for any abelian category $\mathfrak{A}$ there is a natural functor $\mathfrak{Ht}^+(\mathfrak{A}) \to D^+(\mathfrak{A})$; this is the identity on objects, while if $f \in \mathrm{Hom}_{\mathfrak{Ht}^+(\mathfrak{A})}(a^\bullet, b^\bullet)$, one sets $F(f) = \{a^\bullet \xleftarrow{\mathrm{id}} a^\bullet \xrightarrow{\bar{f}} b^\bullet\}$, where $\bar{f}$ is a morphism of complexes in the equivalence class f. One can check that if $a^\bullet \mathrel{\substack{f \\ \rightrightarrows \\ g}} b^\bullet$ are homotopic morphisms of complexes, then the diagrams

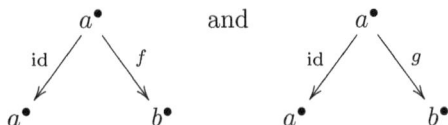

and

determine the same morphism in $D^+(\mathfrak{A})$.

Let us now assume that $\mathfrak{A}$ has enough injectives, and denote by $\mathfrak{I}$ the full subcategory of $\mathfrak{A}$ whose objects are its injective objects. Then one has [36, Proposition 1.7.10] the following proposition.

Proposition 6.4. *The composition of functors* $\mathfrak{Ht}^+(\mathfrak{I}) \to \mathfrak{Ht}^+(\mathfrak{A}) \to D^+(\mathfrak{A})$ *is an equivalence of categories.*

The basis of this result is, of course, the fact that every object in $\mathfrak{A}$ has an injective resolution, and two different injective resolutions are homotopic.

---

[a] For the notion of pullback, see Appendix A.1.

Let $\mathfrak{A}$ and $\mathfrak{B}$ be abelian categories, assume $\mathfrak{A}$ has enough injectives, and let $F\colon \mathfrak{A} \to \mathfrak{B}$ be a left exact functor. It defines a functor

$$\mathfrak{Ht}^+(F)\colon \mathfrak{Ht}^+(\mathfrak{I}) \to \mathfrak{Ht}^+(\mathfrak{B})$$

(simply, a complex of injectives $a^\bullet$ is mapped to the complex $F(a^\bullet)$). In view of Proposition 6.4, the composition

$$\mathfrak{Ht}^+(\mathfrak{I}) \xrightarrow{\ \mathfrak{Ht}^+(F)\ } \mathfrak{Ht}^+(\mathfrak{B}) \to D^+(\mathfrak{B})$$

may be regarded as a functor

$$\mathbb{R}F\colon D^+(\mathfrak{A}) \to D^+(\mathfrak{B}),$$

which we call the *right derived functor of F*. This generalizes the definition of right derived functor given in Chapter 2 as follows: if a is an object in $\mathfrak{A}$, regarded as a complex concentrated in degree 0, then

$$R^i F(a) \simeq H^i(\mathbb{R}F(a)),$$

i.e., the ith right derived functor $R^i F$ applied to the object a yields an object which is isomorphic to the ith cohomology object of the complex $\mathbb{R}F(a)$.

Derived categories are definitely a complicated gadget, but their use may simplify the formulation of results and their proofs very much. As an instance of this fact we reformulate Theorem 5.28 in the language of derived categories.

Theorem 6.5. *Consider a composition of left exact functors*

$$\mathfrak{A} \xrightarrow{\ F\ } \mathfrak{B} \xrightarrow{\ G\ } \mathfrak{C},$$

where $\mathfrak{A}$, $\mathfrak{B}$ and $\mathfrak{C}$ are abelian categories, and $\mathfrak{A}$ and $\mathfrak{B}$ have enough injectives. Suppose that F maps injective objects of $\mathfrak{A}$ to G-acyclic objects of $\mathfrak{B}$. Then $\mathbb{R}(G \circ F)$ and $\mathbb{R}G \circ \mathbb{R}F$ are isomorphic as functors $D^+(\mathfrak{A}) \to D^+(\mathfrak{C})$.

For a proof, see [4, Proposition A.63], among others.

Example 6.6. Two rings with unity, R and S, are said to be *Morita equivalent* if the categories of (say, left) modules R-**mod** and S-**mod** are equivalent. A generalized Morita equivalence may be given in terms of derived categories, i.e., R and S are *derived Morita equivalent* if the derived categories $D(R$-**mod**$)$ and $D(S$-**mod**$)$ are equivalent (as triangulated categories). See [52] for an introduction to this theory.

Example 6.7. Let X be an algebraic variety (see Remark 4.59). The category $\mathfrak{Coh}(X)$ of coherent sheaves of $\mathcal{O}_X$-modules is abelian. We shall denote by $D(X)$ the derived category of complexes of $\mathcal{O}_X$-modules whose cohomology sheaves are coherent. Its subcategory $D^+(X)$ (the derived category of bounded below complexes of $\mathcal{O}_X$-modules whose cohomology sheaves are coherent) is isomorphic to the derived category $D^+(\mathfrak{Coh}(X))$.[b] One can also consider the category $D^b(X)$ of bounded complexes of coherent sheaves (i.e., considering complexes that are bounded both above and below).

There is a theorem, due to Gabriel and Rosenberg [19, 53], according to which a scheme X can be reconstructed from the category $\mathfrak{Qco}(X)$ of quasi-coherent $\mathcal{O}_X$-modules.[c] So in a sense the category $\mathfrak{Qco}(X)$ is "too fine an invariant" of the variety X. The bounded derived category $D^b(X)$ is a more useful invariant; while in some cases it still completely specifies the variety (for instance when X has an ample canonical or anticanonical bundle [4, 6]), in other situations (for instance three-dimensional Calabi–Yau manifolds), its relation with the geometry of X is subtler, and indeed can be used, for instance, to study the birational geometry of X.[d]

On a related note, derived categories can be used to study resolutions of singularities, in particular quotient singularities, or singular varieties that can be desingularized by *small resolutions*.[e] For more information about the use of derived categories to study the geometry of varieties, the reader may refer to [4, 34].

---

[b] For subtleties about the definition of the derived categories $D(X)$, $D^+(X)$, see [4, Appendix A.4.2].

[c] The theorem is originally due to Gabriel, who proved it under the assumption that X is noetherian, while Rosenberg subsequently removed that assumption.

[d] Two irreducible varieties X and Y are said to be birational if there are nonempty open subsets $U \subset X$, $V \subset Y$ that are isomorphic. Note that U and V are dense in X and Y, respectively.

[e] An isolated singularity in a variety is called small if it has a resolution, called *small resolution*, obtained by replacing the singular point with a one-dimensional subvariety.

Appendix

This appendix collects some algebraic results that are used throughout the text, providing complete proofs. The arguments treated are pushouts and pullbacks, the Snake Lemma, and the application of Baer's Criterion to prove that an abelian group is injective if and only if it is divisible.

A.1. Pushouts and Pullbacks

The operation of pushout (also called *fibred coproduct*, *fibred sum*, *co-cartesian square* or *amalgamated sum*) is, loosely speaking, "the most general way" of completing a diagram

$$a \longrightarrow b$$
$$\downarrow$$
$$c$$

in a given category $\mathfrak{C}$. "The most general way" means that it satisfies a suitable universal property. There is also a "dual" notion, called *pullback* (or *fibred product* or *cartesian square*), which is about completing a diagram of the form

$$b$$
$$\downarrow$$
$$c \longrightarrow q$$

Pushouts provide the correct notion of "sum" for morphisms $a \xrightarrow{f} b$, where a is a fixed object in $\mathfrak{C}$, while b and the morphism f vary; and in the same way, pullbacks are the correct notion of "product" for morphisms $b \xrightarrow{g} q$, where q is fixed, and b and g vary.

Remark A.1. A pushout diagram in $\mathfrak{C}$ gives rise to a pullback diagram in the opposite category $\mathfrak{C}^{\mathrm{op}}$, and vice versa, so that proofs holding for pushouts produce dual proofs for pullbacks, and the converse.

Definition A.2. A *pushout* of two morphisms $g : a \to b$, $f : a \to c$ in a category $\mathfrak{C}$ is an object p in $\mathfrak{C}$ with morphisms $h : b \to p$, $k : c \to p$ such that

- the diagram

$$
\begin{array}{ccc}
a & \xrightarrow{g} & b \\
{\scriptstyle f}\downarrow & & \downarrow{\scriptstyle h} \\
c & \xrightarrow{k} & p
\end{array}
\tag{A.1}
$$

commutes;
- the triple (p, h, k) is universal with respect to this property, i.e., if (p', h', k') is another triple satisfying the same property, there is a morphism $u : p \to p'$ such that the diagram

$$
\tag{A.2}
$$

commutes.

By definition, the triple (p, h, k) is unique up to unique isomorphism. The pushout of b and c over a is denoted

$$
b \oplus_a c.
$$

This notation is motivated by the following (very useful) fact.

Lemma A.3. *If* $\mathfrak{A}$ *is an abelian category, the object* $p = (b \oplus c)/\operatorname{im}(g, -f)$, *with the morphisms* h *and* k *given by the natural inclusions into the direct sum followed by the projection onto the quotient, is a pushout of* g *and* f.

If $\mathfrak{A}$ is a concrete category,[a] $\text{im}(g, -f)$ is the subobject of $a \oplus b$ formed by elements of the type $(g(x), -f(x))$ for some $x \in a$. In general, the sequence

$$a \xrightarrow{(g,-f)} b \oplus c \xrightarrow{(h,k)} p \to 0 \qquad (A.3)$$

is exact.

Proof. Set $p = (b \oplus c)/\text{im}(g, -f)$. If p' is another pushout as in diagram (A.2), note that the morphism $h' + k' : b \oplus c \to p'$ vanishes on $\text{im}(g, -f)$, so that it provides the required morphism $u : p \to p'$.

If $\mathfrak{A}$ is a concrete category, we can write the proof as follows. If $z = [(x, y)] \in p$, define u as $u(z) = h'(x) + k'(y)$. This is well defined because if also $z = [(x', y')]$, then

$$x' = x + g(t), \quad y' = y - f(t)$$

for some $t \in a$, so that

$$h'(x') + k'(y') = h'(x) + h'(g(t)) + k'(y) - k'(f(t)) = h'(x) + k'(y). \quad \square$$

The notion of *pullback* is dual to that of pushout.

Definition A.4. The pullback of two morphisms $h : b \to q$ and $k : c \to q$ in a category $\mathfrak{C}$ is an object p with morphisms $g : p \to b$, $f : p \to c$ such that

• the diagram

$$
\begin{array}{ccc}
p & \xrightarrow{g} & b \\
{\scriptstyle f}\downarrow & & \downarrow{\scriptstyle h} \\
c & \xrightarrow{k} & q
\end{array}
\qquad (A.4)
$$

commutes;
• the triple (p, g, f) is universal with respect to this property, i.e., if (p', f', g') is another triple satisfying the same property, there is a morphism $v : p' \to p$ such that the diagram

commutes.

---

[a]The notion of concrete category was given in the footnote d in Chapter 1.

(We already introduced this construction at p. 70 for the category of topological spaces, and at p. 107 for the category of schemes, calling it *fibred product*.)

Again, pullbacks are unique up to unique isomorphism. The pullback of b and c over q is denoted

$$b \times_q c.$$

If $\mathfrak{A}$ is an abelian category, one can define the pullback as

$$p = \ker(h, -k) : b \oplus c \to q. \tag{A.5}$$

Exercise A.5. Prove, in analogy with the proof of Lemma A.3, that when $\mathfrak{A}$ is an abelian category, equation (A.5) defines a pullback.

So for an abelian category the diagram (A.4) is a pullback if and only if the sequence

$$0 \to p \xrightarrow{(g,f)} b \oplus c \xrightarrow{(h,-k)} q \tag{A.6}$$

is exact.

Lemma A.6. *Let $\mathfrak{A}$ be an abelian category. If the diagram (A.4) is a pullback, then $\ker k \simeq \ker g$. Analogously, if the diagram (A.1) is a pushout, then $\operatorname{coker} g \simeq \operatorname{coker} k$.*

Proof. We prove the first claim. Let us consider the diagram

$$
\begin{array}{ccccccc}
0 & \longrightarrow & \ker g & \xrightarrow{i} & p & \xrightarrow{g} & b \\
& & \downarrow{\scriptstyle e} & & \downarrow{\scriptstyle f} & & \downarrow{\scriptstyle h} \\
0 & \longrightarrow & \ker k & \xrightarrow{j} & c & \xrightarrow{k} & q
\end{array}
$$

where the square on the right is a pullback, and e is the morphism induced by f. Since the sequence (A.6) is exact, from the diagram

$$
\begin{array}{ccc}
& & \ker k \\
& {\scriptstyle t}\nearrow\!\!\!\!\dashrightarrow & \downarrow{\scriptstyle (0,j)} \\
0 \longrightarrow & p \xrightarrow{(g,f)} b \oplus c & \xrightarrow{(h,-k)} q
\end{array}
$$

and from the universal property of the kernel (note that $(h, -k) \circ (0, j) = 0$) we have the existence of a morphism $t : \ker k \to p$ such that $g \circ t = 0$; by the same reasoning, using the diagram

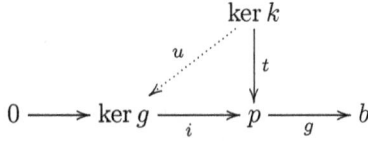

$$
\begin{array}{ccccccc}
 & & & & \ker k & & \\
 & & & \swarrow{\scriptstyle u} & \downarrow{\scriptstyle t} & & \\
0 & \longrightarrow & \ker g & \xrightarrow{\ i\ } & p & \xrightarrow{\ g\ } & b
\end{array}
$$

there is a morphism $u : \ker k \to \ker g$ such that $t = i \circ u$. Note that one also has $f \circ t = j$. From the diagram, one sees that

$$i \circ u \circ e = i, \quad j \circ e \circ u = f \circ i \circ u = f \circ t = j.$$

As i and j are both monomorphisms, we have

$$u \circ e = \mathrm{id}_{\ker g}, \quad e \circ u = \mathrm{id}_{\ker k}$$

so that the first claim follows. The second claim is proved in a dual way. $\qquad \square$

Lemma A.7. *Let $\mathfrak{A}$ be an abelian category. If the diagram* (A.1) *is a pushout, and g is monomorphism, then the diagram is also a pullback, and moreover k is a monomorphism as well.*

Analogously, if the diagram (A.4) *is a pullback, and the morphism k is an epimorphism, then the diagram is also a pushout, and g is an epimorphism as well.*

Proof. We prove the first claim. Let us check that the diagram (A.1) is a pullback. We know that the sequence (A.3) is exact. On the other hand, as g is injective, also $(f, -g)$ is injective, so that the sequence (A.6) is exact, and therefore (A.1) is a pullback.

Now we consider the diagram

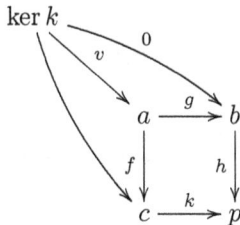

$$
\begin{array}{ccc}
\ker k & & \\
 & \searrow^{\scriptstyle 0} & \\
{\scriptstyle v} & a \xrightarrow{\ g\ } b & \\
 & f\downarrow \quad \downarrow h & \\
 & c \xrightarrow{\ k\ } p &
\end{array}
$$

where the morphism v, given by the universal property of the pullback, is a monomorphism as $\ker k$ injects into c. Since g is monomorphism as well, and $g \circ v = 0$, one has $\ker k = 0$.

The second claim is proved in a dual way. ☐

The following result will be used in the Appendix A.2.

Lemma A.8. *Let* $p \xrightarrow{f} d \xrightarrow{g} q$ *be a complex in an abelian category. The following conditions are equivalent:*

- *the sequence* $p \xrightarrow{f} d \xrightarrow{g} q$ *is exact;*
- *for any morphism* $h : s \to d$ *such that* $g \circ h = 0$, *there exist an epimorphism* $f' : s' \twoheadrightarrow s$ *and a commutative diagram*

$$
\begin{array}{ccc}
s' & \xrightarrow{\ f'\ } & s \\
\downarrow & \underset{h}{\Big\downarrow} & \Big\downarrow{\scriptstyle 0} \\
p & \xrightarrow[f]{} d \xrightarrow[g]{} & q
\end{array}
$$

Proof. If the complex is exact, let $s' = p \times_{\ker g} s$. Since $p \to \ker g$ is an epimorphism, $s' \to s$ is an epimorphism by Lemma A.7. The converse is proven choosing $s = \ker g$. Indeed, the composition $s' \to p \to \ker g$ is an epimorphism. Hence $p \to \ker g$ is an epimorphism. ☐

For more results on pushouts and pullbacks, the reader may consult [36, 49, Section 12.5], or [48, especially Propositions 7.1 and 20.2], and [30].

A.2. Snake Lemma

The Snake Lemma is a standard trick in homological algebra, which for instance is used to construct long exact sequences in homology or cohomology. One starts with a commutative diagram of morphisms in an abelian category $\mathfrak{A}$

$$
\begin{array}{ccccccc}
p & \xrightarrow{f} & d & \xrightarrow{g} & q & \longrightarrow & 0 \\
\Big\downarrow{\scriptstyle u} & & \Big\downarrow{\scriptstyle v} & & \Big\downarrow{\scriptstyle w} & & \\
0 & \longrightarrow & p' & \xrightarrow[f']{} & d' & \xrightarrow[g']{} & q'
\end{array}
\tag{A.7}
$$

whose rows are exact. We complete the diagram (A.7) adding kernels and cokernels:

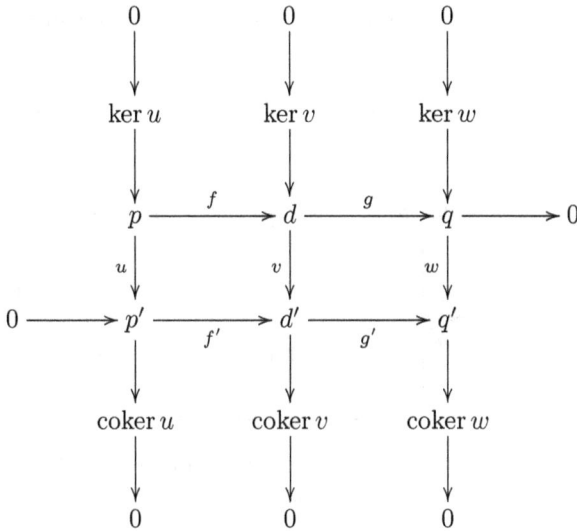

$$
\begin{array}{ccccc}
& 0 & & 0 & & 0 \\
& \downarrow & & \downarrow & & \downarrow \\
& \ker u & & \ker v & & \ker w \\
& \downarrow & & \downarrow & & \downarrow \\
& p \xrightarrow{\ f\ } & d \xrightarrow{\ g\ } & q \longrightarrow 0 \\
u\downarrow & & v\downarrow & & w\downarrow \\
0 \longrightarrow & p' \xrightarrow[\ f'\ ]{} & d' \xrightarrow[\ g'\ ]{} & q' \\
& \downarrow & & \downarrow & & \downarrow \\
& \operatorname{coker} u & & \operatorname{coker} v & & \operatorname{coker} w \\
& \downarrow & & \downarrow & & \downarrow \\
& 0 & & 0 & & 0
\end{array}
$$

It is easy to see that one can define morphisms

$$\ker u \xrightarrow{\ \bar{f}\ } \ker v \xrightarrow{\ \bar{g}\ } \ker w$$

and

$$\operatorname{coker} u \xrightarrow{\ \bar{f}'\ } \operatorname{coker} v \xrightarrow{\ \bar{g}'\ } \operatorname{coker} w$$

and the diagram completed with them commutes. The whole point of the Snake Lemma is the existence of a morphism $\ker w \to \operatorname{coker} u$ such that the three kernels and the three cokernels fit into an exact sequence.

Proposition A.9 (Snake Lemma). *There exists a morphism $\delta : \ker w \to \operatorname{coker} u$ such that the sequence*

$$\ker u \xrightarrow{\ \bar{f}\ } \ker v \xrightarrow{\ \bar{g}\ } \ker w \xrightarrow{\ \delta\ } \operatorname{coker} u \xrightarrow{\ \bar{f}'\ } \operatorname{coker} v \xrightarrow{\ \bar{g}'\ } \operatorname{coker} w \qquad (A.8)$$

is exact. Moreover,

- *if f is a monomorphism, $\bar{f}$ is a monomorphism as well;*
- *if g' is epimorphism, $\bar{g}'$ is epimorphism as well.*

Proof. We follow the proof given in [36] adding some details.

Step 1. We define $\delta : \ker w \to \operatorname{coker} u$. Let $W = d \times_q \ker w$ and $Z = d' \oplus_{p'} \operatorname{coker} u$; denote $h : W \to \ker w$ the natural morphism. One has a commutative diagram

The morphism h is an epimorphism due to Lemma A.7. We prove that the composition $W \to d \xrightarrow{v} d' \to Z$ factors as

$$W \to \ker w \xrightarrow{\delta} \operatorname{coker} u \hookrightarrow Z.$$

In fact, the composition $W \to d' \to q'$ vanishes, so that the morphism $W \to d'$ factors through p'. The morphism $\ker h \to \ker g$ is an isomorphism by Lemma A.6. Note that $p \to \ker g$ is an epimorphism, so that $\ker g \to p' \to \operatorname{coker} u$ vanishes. Therefore, the composition $\ker h \to W \to \operatorname{coker} u$ vanishes; noting that $\ker w \simeq \operatorname{coker}(\ker h \to W)$, by the universal property of the cokernels the morphism $W \to \operatorname{coker} u$ factors as $W \twoheadrightarrow \ker w \xrightarrow{\delta} \operatorname{coker} u$ as we see from the diagram

Step 2. The sequence

$$\ker u \xrightarrow{\bar{f}} \ker v \xrightarrow{\bar{g}} \ker w \xrightarrow{\delta} \operatorname{coker} u \xrightarrow{\bar{f}'} \operatorname{coker} v \xrightarrow{\bar{g}'} \operatorname{coker} w$$

is exact.

We start by proving the exactness of $\ker v \xrightarrow{\bar{g}} \ker w \xrightarrow{\delta} \operatorname{coker} u$. We shall use Lemma A.8; so let $\phi : s \to \ker w$ be a morphism such that $\delta \circ \phi = 0$ (a pair (s, ϕ) with this property always exists, for instance take $s = \ker \delta$). Since $W \to \ker w$ is an epimorphism, if we define $s_1 = s \times_{\ker w} W$

by Lemma A.7, $s_1 \twoheadrightarrow s$ is an epimorphism as well. Splicing the resulting pullback diagram with the left square in (A.7), we obtain

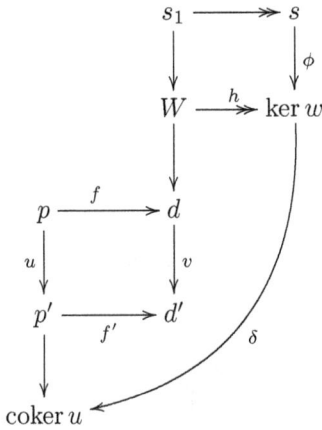

In diagram (A.2) we defined a morphism $W \to p'$. Now the composition $s_1 \to W \to p' \to \operatorname{coker} u$ is the same as the composition $s_1 \to s \to \ker w \to \operatorname{coker} u$, which is zero, so that we can draw the diagram

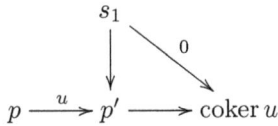

By Lemma A.8, there exists an epimorphism $s_0 \twoheadrightarrow s_1$ such that the composition $s_0 \to s_1 \to W \to p'$ decomposes as $s_0 \xrightarrow{k} p \xrightarrow{u} p'$ (note that one can take $s_0 = s_1 \times_{p'} p$). Denote by λ the composition $s_0 \to s_1 \to W \to d$. At this stage, we have a diagram

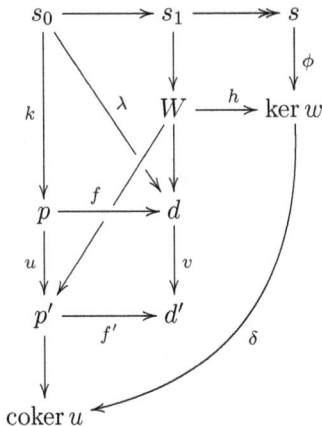

Now $v \circ \lambda = v \circ f \circ k$, as follows from the commutativity of the diagram

$$
\begin{array}{ccc}
s_0 & \longrightarrow & s_1 \\
\downarrow{\scriptstyle k} & & \downarrow \\
p & \xrightarrow{u} & p' \\
\downarrow{\scriptstyle f} & & \downarrow{\scriptstyle f'} \\
d & \xrightarrow{v} & d'
\end{array}
$$

Therefore, $\lambda - f \circ k : s_0 \to p$ factors through $\ker v$, and we obtain a commutative diagram

$$
\begin{array}{ccc}
s_0 & \longrightarrow & s \\
{\scriptstyle \lambda - f \circ k}\downarrow & & \downarrow \quad \searrow^{0} \\
\ker v & \longrightarrow & \ker w \longrightarrow \operatorname{coker} u \\
\uparrow & & \uparrow \\
d & \xrightarrow{g} & q
\end{array}
$$

Finally, one applies Lemma A.8 again.

One proves that the sequences $\ker u \xrightarrow{\bar{f}} \ker v \xrightarrow{\bar{g}} \ker w$ and $\operatorname{coker} u \xrightarrow{\bar{f}'} \operatorname{coker} v \xrightarrow{\bar{g}'} \operatorname{coker} w$ are exact by similar arguments, again using Lemma A.8 repeatedly. This proves the first claim.

To prove the second claim, we note that, when f is a monomorphism, $\ker u \to \ker v$ is the restriction of a monomorphism; the third claim is proved by a similar reasoning, i.e., the morphism $\operatorname{coker} v \to \operatorname{coker} w$ is the quotient of an epimorphism. $\qquad\qquad\square$

If $\mathfrak{A}$ is a concrete category, there is a much simpler proof using diagram chasing. In particular, the morphism δ is defined as follows. If x is an element in $\ker w$, push it to q, and take an inverse image in d; now apply v. What one gets is in the kernel of g', and therefore lies in p', so that we can push it down to $\operatorname{coker} u$. This is well defined because the difference of two different counterimages via g lies in p, so that it is zero in $\operatorname{coker} u$. The exactness at each step of the sequence (A.8) is vey simply, albeit tediously, proved explicitly. This in particular works for the category of R-modules, where R is a ring. From here one can get the result for a general *small* abelian

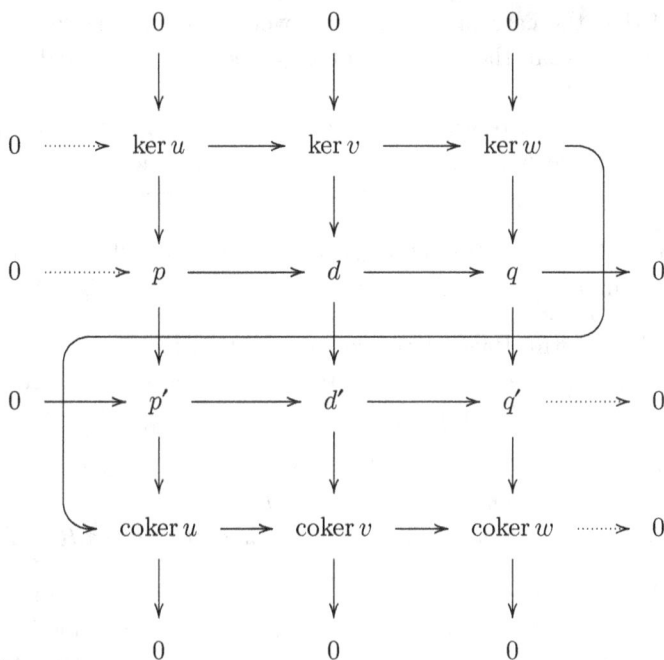

Figure A.1.   The diagram describing the Snake Lemma.

category (i.e., an abelian category whose class of objects and all Hom classes are sets) by using the Freyd–Mitchell Embedding Theorem [17, 47].[b]

Theorem A.10. *If $\mathfrak{A}$ is a small abelian category, there exists a ring R with unity and an exact fully faithful functor $\mathfrak{A} \to R$-**mod**.*

The final situation is depicted in the diagram in Figure A.1. The shape of the arrow representing the morphism $\delta : \ker w \to \operatorname{coker} u$ motivates the name "Snake Lemma".

A.3. Baer's Criterion

As we discussed in Section 2.1, the injective abelian groups are the divisible groups, i.e., groups such that for every integer and every element g in the group, there is another element h such that $g = nh$. This fact is the key to

---

[b]In particular, this implies that any small abelian category is concretizable, see footnote d in Chapter 1. Actually, all small categories are concretizable.

the proof that the category of modules over a ring has enough injectives, and eventually that the category of $\mathcal{O}_X$-modules has enough injectives as well.

More generally, here we consider a commutative ring with unity R and study the divisible R-modules. We state and prove Baer's Criterion, i.e., a criterion for checking whether an R-module is injective.

Definition A.11. An R-module M is divisible if, for all $r \in R$, not a zero divisor, the multiplication by it defines a surjective morphism $M \to M$.

In its basic form, Baer's Criterion states that, to check whether an R-module M is injective, it is enough to consider the action of the functor $\text{Hom}_R(-, M)$ on the ideals of R.

Proposition A.12 (Baer's Criterion). *An R-module I is injective if and only if every homomorphism of R-modules $\mathfrak{s} \to I$, where $\mathfrak{s} \subset R$ is any ideal of R, can be extended to a homomorphism of modules $R \to I$.*

Proof. We only need to prove the "if" part because the other is the definition of injective object. Let N be a submodule of an R-module M and consider a homomorphism $f : N \to I$. We prove that it can be extended to M assuming the hypothesis. The set

$$W = \{(N', f') \text{ where } N \subset N' \subset M \text{ and } f' : N' \to I \text{ is an extension of } f\}$$

is partially ordered by $f' \leq f''$ if and only if $N' \subset N''$ and f'' extends f'. Using Zorn's Lemma we may pick up a maximal R-module $\bar{N}$ and a maximal extension $\bar{f} : \bar{N} \to I$.

Suppose $\bar{N} \neq M$; then there exists $m \in M - \bar{N}$ and $\bar{N} + Rm$ is a submodule of M. Moreover, we claim that there exists an extension of $\bar{f}$, say $g : \bar{N} + nR \to I$, which proves that $\bar{N} = M$. Indeed, the ideal $\mathfrak{s} = \{s \in R \mid sm \in \bar{N}\}$ of R gives rise a homomorphism

$$\mathfrak{s} \to \bar{N} \to I,$$

$$s \mapsto sm \mapsto \bar{f}(sm),$$

which by hypothesis extends to a morphism $\rho : R \to I$. Then we define the homomorphism

$$g : \bar{N} + Rm \to I,$$

$$n + rm \mapsto \bar{f}(n) + \rho(r).$$

$\square$

Proposition A.13. *Injective R-modules are divisible.*

Proof. Let M be an injective R-module, and let $m \in M$, $r \in R$. Consider the diagram

$$0 \longrightarrow \langle r \rangle \longrightarrow R$$
$$\downarrow \qquad \qquad g$$
$$M$$

where $f : \langle r \rangle \to M$ is the morphism which maps r to m. Since M is injective, f extends to a morphism $g : R \to M$; then $m = rg(1)$, so that the morphism $M \to M$ given by the multiplication by r is surjective. $\square$

The converse, i.e., to show that divisible modules are injective, is a more delicate matter. For principal ideal domains, this follows from Baer's Criterion.

Corollary A.14. *If R is a principal ideal domain, a divisible R-module is injective.*

Proof. Let M be a divisible R-module. For any ideal $\mathfrak{r}$ of R, there exists $r \in R$ such that $\langle r \rangle = \mathfrak{r}$. Therefore, for any $m \in M$, the morphism $\rho : \mathfrak{r} \to M$ sending r to m can be extended to $\bar{\rho} : R \to M$ sending $1 \mapsto m'$ (as M is divisible there exists $m' \in M$ such that $rm' = m$). By Baer's Criterion this proves that M is injective. $\square$

In particular, all abelian groups G, i.e., $\mathbb{Z}$-modules, are injective if and only if they are divisible.

A.4. Additional Exercises

1. Let

$$0 \to M \to N \to P \to 0$$

be an exact sequence of modules over a ring R. Prove that if $f : M \to M'$ is a homomorphism of R-modules then there exists an exact sequence

$$0 \to M' \to N' \to P \to 0$$

for some R-module N'.

2. Let M, N, M', N' be modules over a ring R and let

$$
\begin{array}{ccc}
M & \xrightarrow{\ f\ } & N \\
\downarrow{\scriptstyle u} & & \downarrow{\scriptstyle v} \\
M' & \xrightarrow[\ g\ ]{} & N'
\end{array}
$$

be a commutative diagram, where g is a monomorphism. Prove that there exist an R-module K and morphisms $\ker v \to K$, $K \to \operatorname{coker} u$ such that the sequence

$$\ker u \xrightarrow{\ \bar f\ } \ker v \to K \to \operatorname{coker} u \xrightarrow{\ \bar g\ } \operatorname{coker} v$$

is exact, where $\bar f$ and $\bar g$ are the morphisms induced by f and g, respectively.

3. Let $0 \to a \to b \to c \to 0$ and $0 \to a' \to b' \to c \to 0$ be two exact sequences in an abelian category. Use the pullback construction and (possibly) the Snake Lemma to build the diagram

$$
\begin{array}{ccccccccc}
& & 0 & & 0 & & & & \\
& & \downarrow & & \downarrow & & & & \\
& & a' & = = & a' & & & & \\
& & \downarrow & & \downarrow & & & & \\
0 & \to & a & \to & d & \to & b' & \to & 0 \\
& & \| & & \downarrow & & \downarrow & & \\
0 & \to & a & \to & b & \to & c & \to & 0 \\
& & & & \downarrow & & \downarrow & & \\
& & & & 0 & & 0 & &
\end{array}
$$

4. Let $0 \to a' \to a \to a'' \to 0$ and $0 \to a' \to d \to b \to 0$ be two exact sequences in an abelian category. Use the pushout construction and (possibly) the Snake Lemma to build the diagram

$$
\begin{array}{ccc}
0 & & 0 \\
\downarrow & & \downarrow \\
0 \longrightarrow a' \longrightarrow d \longrightarrow b \longrightarrow 0 \\
\downarrow & & \downarrow \quad \parallel \\
0 \longrightarrow a \longrightarrow d' \longrightarrow b \longrightarrow 0 \\
\downarrow & & \downarrow \\
a'' = \!\!= a'' \\
\downarrow & & \downarrow \\
0 & & 0
\end{array}
$$

5. Consider a commutative diagram in an abelian category $\mathfrak{A}$

$$
\begin{array}{ccccc}
a & \xrightarrow{f} & b & \xrightarrow{g} & c \\
\downarrow{\scriptstyle u} & & \downarrow{\scriptstyle v} & & \downarrow{\scriptstyle w} \\
a' & \xrightarrow{f'} & b' & \xrightarrow{g'} & c'
\end{array}
$$

Prove the following statements:

(a) if the first line of the diagram is exact and f' is a monomorphism, the sequence

$$\ker u \xrightarrow{f} \ker v \xrightarrow{g} \ker w$$

is exact;

(b) if the second line is exact and g is an epimorphism, the sequence

$$\operatorname{coker} u \xrightarrow{\bar{f}'} \operatorname{coker} v \xrightarrow{\bar{g}'} \operatorname{coker} w$$

is exact (a bar denotes the morphisms induced on the quotients).

Hint: you may like to solve the exercise by first assuming that $\mathfrak{A}$ is a category of modules over a ring, and then generalizing to any abelian category.

Bibliography

[1] M. F. Atiyah and I. G. Macdonald, *Introduction to Commutative Algebra*, Addison-Wesley Series in Mathematics, Westview Press, Boulder, CO, economy ed., 2016.

[2] W. P. Barth, K. Hulek, C. A. M. Peters, and A. Van de Ven, *Compact Complex Surfaces*, Ergebnisse der Mathematik und ihrer Grenzgebiete. 3. Folge. A Series of Modern Surveys in Mathematics, Vol. 4, Springer-Verlag, Berlin, second ed., 2004.

[3] C. Bartocci, U. Bruzzo, and D. Hernández Ruipérez, *The Geometry of Supermanifolds*, Mathematics and its Applications, Vol. 71, Kluwer Academic Publishers Group, Dordrecht, 1991.

[4] ———, *Fourier–Mukai and Nahm Transforms in Geometry and Mathematical Physics*, Progress in Mathematics, Vol. 276, Birkhäuser Boston, Inc., Boston, MA, 2009.

[5] G. M. Bergman, *An Invitation to General Algebra and Universal Constructions*, Universitext, Springer, Cham, second ed., 2015.

[6] A. Bondal and D. Orlov, Reconstruction of a variety from the derived category and groups of autoequivalences, *Compositio Math.*, 125 (2001), pp. 327–344.

[7] W. M. Boothby, *An Introduction to Differentiable Manifolds and Riemannian Geometry*, Pure and Applied Mathematics, Vol. 120, Academic Press, Inc., Orlando, FL, second ed., 1986.

[8] R. Bott and L. W. Tu, *Differential Forms in Algebraic Topology*, Graduate Texts in Mathematics, Vol. 82, Springer-Verlag, Berlin, 1982.

[9] N. Bourbaki, *Algebra. I. Chapters 1–3*, Elements of Mathematics (Berlin), Springer-Verlag, Berlin, 1989. Translated from the French, Reprint of the 1974 edition.

[10] ———, *Commutative Algebra. Chapters 1–7*, Elements of Mathematics (Berlin), Springer-Verlag, Berlin, 1998. Translated from the French, Reprint of the 1989 English translation.

[11] G. E. Bredon, *Sheaf Theory*, Graduate Texts in Mathematics, Vol. 170, Springer-Verlag, New York, second ed., 1997.

[12] J. S. Calcut and J. D. McCarthy, *Topological Pullback, Covering Spaces, and a Triad of Quillen*, preprint, 2012, arXiv:1205.3122 [math.GN].

[13] C. Chevalley and S. Eilenberg, Cohomology theory of Lie groups and Lie algebras, *Trans. Amer. Math. Soc.*, 63 (1948), pp. 85–124.

[14] G. De Rham, Sur l'analysis situs des variétés à n dimensions, 1931. http://www.numdam.org/item/THESE_1931__129__1_0/.

[15] A. Dold, *Lectures on Algebraic Topology*, Springer Classics in Mathematics, Springer-Verlag, Berlin, 1980.

[16] D. Eisenbud, *Commutative Algebra with a View Toward Algebraic Geometry*, Graduate Texts in Mathematics, Vol. 150, Springer-Verlag, New York, 1995.

[17] P. Freyd, *Abelian Categories. An Introduction to the Theory of Functors*, Harper's Series in Modern Mathematics, Harper & Row Publishers, New York, 1964.

[18] ——, *On the Concreteness of Certain Categories*, Symposia Mathematica, Vol. 4 (INDAM, Rome, 1968/69), Academic Press, London, 1970, pp. 431–456.

[19] P. Gabriel, Des catégories abéliennes, *Bull. Soc. Math. France*, 90 (1962), pp. 323–448.

[20] R. Godement, *Topologie Algébrique et Théorie des Faisceaux*, Hermann, Paris, 1973. Troisième édition revue et corrigée, Publications de l'Institut de Mathématique de l'Université de Strasbourg, XIII, Actualités Scientifiques et Industrielles, No. 1252.

[21] P. Griffiths and J. Harris, *Principles of Algebraic Geometry*, Wiley Classics Library, John Wiley & Sons, Inc., New York, 1994. Reprint of the 1978 original.

[22] A. Grothendieck, Sur quelques points d'algèbre homologique, *Tôhoku Math. J.* (2), 9 (1957), pp. 119–221.

[23] ——, Éléments de géométrie algébrique. I. Le langage des schémas, *Inst. Hautes Études Sci. Publ. Math.*, (1960), p. 228.

[24] ——, Éléments de géométrie algébrique. II. Étude globale élémentaire de quelques classes de morphismes, *Inst. Hautes Études Sci. Publ. Math.*, (1961), p. 222.

[25] ——, Éléments de géométrie algébrique. III. Étude cohomologique des faisceaux cohérents. I, *Inst. Hautes Études Sci. Publ. Math.*, (1961), p. 167.

[26] ——, Éléments de géométrie algébrique. IV. Étude locale des schémas et des morphismes de schémas IV, *Inst. Hautes Études Sci. Publ. Math.*, (1967), p. 361.

[27] P. R. Halmos, *Naive Set Theory*, Springer-Verlag, New York, 1974. Reprint of the 1960 edition, Undergraduate Texts in Mathematics.

[28] R. Hartshorne, *Algebraic Geometry*, Graduate Texts in Mathematics, Vol. 52, Springer-Verlag, New York, Heidelberg, 1977.

[29] A. Hatcher, *Algebraic Topology*, Cambridge University Press, Cambridge, 2002.

[30] P. J. Hilton and U. Stammbach, *A Course in Homological Algebra*, Graduate Texts in Mathematics, Vol. 4, Springer-Verlag, New York, second ed., 1997.

[31] F. Hirzebruch, *Topological Methods in Algebraic Geometry*, Classics in Mathematics, Springer-Verlag, Berlin, 1995.

[32] G. Hochschild and J.-P. Serre, Cohomology of Lie algebras, *Ann. of Math.* (2), 57 (1953), pp. 591–603.

[33] J. E. Humphreys, *Introduction to Lie Algebras and Representation Theory*, Graduate Texts in Mathematics, Vol. 9, Springer-Verlag, Berlin, 1978. Second printing, revised.

[34] D. Huybrechts, *Fourier–Mukai Transforms in Algebraic Geometry*, Oxford Mathematical Monographs, Oxford University Press, Oxford, 2006.

[35] N. Jacobson, *Lie Algebras*, Dover Publications, Inc., New York, 1979. Republication of the 1962 original.

[36] M. Kashiwara and P. Schapira, *Categories and Sheaves*, Grundlehren der Mathematischen Wissenschaften [Fundamental Principles of Mathematical Sciences], Vol. 332, Springer-Verlag, Berlin, 2006.

[37] J. L. Kelley, *General Topology*, Graduate Texts in Mathematics, Vol. 27. Springer-Verlag, Berlin, 1975. Reprint of the 1955 edition [Van Nostrand, Toronto, ON].

[38] S. Kobayashi, *Differential Geometry of Complex Vector Bundles*, Princeton Legacy Library, Princeton University Press, Princeton, NJ, 2014. Reprint of the 1987 edition.

[39] S. Kobayashi and K. Nomizu, *Foundations of Differential Geometry. Vol. I*, Wiley Classics Library, John Wiley & Sons, Inc., New York, 1996. Reprint of the 1963 original, A Wiley-Interscience Publication.

[40] ———, *Foundations of Differential Geometry. Vol. II*, Wiley Classics Library, John Wiley & Sons, Inc., New York, 1996. Reprint of the 1969 original, A Wiley-Interscience Publication.

[41] S. Lang, *Algebra*, Graduate Texts in Mathematics, Vol. 211, Springer-Verlag, New York, third ed., 2002.

[42] F. W. Lawvere and S. H. Schanuel, *Conceptual Mathematics: A First Introduction to Categories*, Cambridge University Press, Cambridge, second ed., 2009.

[43] S. Mac Lane, *Categories for the Working Mathematician*, Graduate Texts in Mathematics, Vol. 5, Springer-Verlag, New York, second ed., 1998.

[44] W. S. Massey, Exact couples in algebraic topology. I, II, *Ann. of Math.* (2), 56 (1952), pp. 363–396.

[45] H. Matsumura, *Commutative Algebra*, Mathematics Lecture Note Series, Vol. 56, Benjamin/Cummings Publishing Co., Inc., Reading, MA, second ed., 1980.

[46] ———, *Commutative Ring Theory*, Cambridge Studies in Advanced Mathematics, Vol. 8, Cambridge University Press, Cambridge, 1986. Translated from the Japanese by M. Reid.

[47] B. Mitchell, The full embedding theorem, *Am. J. Math.*, 86 (1964), pp. 619–637.

[48] ——, *Theory of Categories*, Pure and Applied Mathematics, Vol. 17, Academic Press, New York, 1965.

[49] Multiple Authors, *The Stacks Project*. https://stacks.math.columbia.edu/.

[50] D. Murfet, *Section 3.2 — Cohomology of Sheaves*. http://therisingsea.org/notes/.

[51] H. Poincaré, Analysis situs, *J. École Polytechnique* (1), 2 (1895), pp. 1–123.

[52] J. Rickard, *Morita theory for derived categories*, J. London Math. Soc. (2), 39 (1989), pp. 436–456.

[53] A. L. Rosenberg, The spectrum of abelian categories and reconstruction of schemes, in *Rings, Hopf Algebras, and Brauer Groups* (Antwerp/Brussels, 1996), Lecture Notes in Pure and Applied Mathematics, Vol. 197, Dekker, New York, 1998, pp. 257–274.

[54] J. J. Rotman, *An Introduction to Homological Algebra*, Universitext, Springer, New York, second ed., 2009.

[55] J.-P. Serre, Faisceaux algébriques cohérents, *Ann. of Math.* (2), 61 (1955), pp. 197–278.

[56] ——, Sur la cohomologie des variétés algébriques, *J. Math. Pures Appl.* (9), 36 (1957), pp. 1–16.

[57] E. H. Spanier, *Algebraic Topology*, Springer-Verlag, New York, 1995. Corrected reprint of the 1966 original.

[58] S. Sternberg, *Lectures on Differential Geometry*, Chelsea Publishing Co., New York, second ed., 1983. With an appendix by Shlomo Sternberg and Victor W. Guillemin.

[59] B. R. Tennison, *Sheaf Theory*, London Mathematical Society Lecture Note Series, Vol. 20. Cambridge University Press, Cambridge, 1975.

[60] J.-L. Verdier, Catégories dérivées: quelques résultats (état 0), in *Cohomologie Étale*, Lecture Notes in Mathematics, Vol. 569, Springer, Berlin, 1977, pp. 262–311. (SGA $4\frac{1}{2}$).

[61] ——, Des catégories dérivées des catégories abéliennes, *Astérisque*, (1996), pp. xii+253 pp. (1997). With a preface by Luc Illusie, Edited and with a note by Georges Maltsiniotis.

[62] C. Voisin, *Hodge Theory and Complex Algebraic Geometry. I*, Cambridge Studies in Advanced Mathematics, Vol. 76, Cambridge University Press, Cambridge, 2002. Translated from the French original by Leila Schneps.

[63] F. W. Warner, *Foundations of Differentiable Manifolds and Lie Groups*, Graduate Texts in Mathematics, Vol. 94, Springer-Verlag, Berlin, 1983. Corrected reprint of the 1971 edition.

[64] C. A. Weibel, *An Introduction to Homological Algebra*, Cambridge Studies in Advanced Mathematics, Vol. 38, Cambridge University Press, Cambridge, 1994.

[65] R. O. Wells, Jr., *Differential Analysis on Complex Manifolds*, Graduate Texts in Mathematics, Vol. 65, Springer, New York, third ed., 2008. With a new appendix by Oscar García-Prada.

[66] H. Whitney, On products in a complex, *Ann. of Math.* (2), 39 (1938), pp. 397–432.

Index